ÖSTERREICHISCHE AKADEMIE DER WISSENSCHAFTEN
PHILOSOPHISCH-HISTORISCHE KLASSE
**SITZUNGSBERICHTE**, 512. BAND

OTTO NEUGEBAUER

# CHRONOGRAPHY IN ETHIOPIC SOURCES

VERLAG DER
ÖSTERREICHISCHEN AKADEMIE DER WISSENSCHAFTEN

WIEN 1989

Vorgelegt vom Sekretär WERNER WELZIG
in der Sitzung am 14. Oktober 1987


— ISBN 3 7001 1469 9

Herstellung: Druckerei G. Grasl, A-2540 Bad Vöslau

# PREFACE

Einen letzten Strich
tat der Geigerich —
dann war nichts weiter
zu beweisen.

Morgenstern, Galgenlieder

In a previous study on Ethiopic Computus-texts I was concerned with calendrical problems, mainly Easter. But the same source material — some hundred unpublished manuscripts — frequently contains also chronological lists, ranging from Biblical events to Ethiopic kings. Analyzing these texts one by one, I assembled some 80 tabulations which could, I hope, be of some interest to scholars engaged in medieval studies. For Ethiopic chronology, they provide insight in the important role played by the 532-year cycle, in particular for the double-dates assigned to Yekuno Amlâk and hence to Ethiopic chronology in general (p. 57).

An unexpected reference to the legend of Alexander the "Bicornute" is unfortunately only fragmentarily preserved (p. 46). The chronology of the Kingdom of Aksum can be brought, by a simple emendation, into an at least numerically consistent form (p. 59). If on the whole the material presented here is not very startling, it seems to me nevertheless better to present it in a controllable fashion than to ignore it.

O. N.

## CONTENTS

# I. INTRODUCTION

## 1. Notations

The texts used in this study are the same as listed in EAC p. 245 to 255. Some abbreviations used in AEC are here further shortened:

| | | | |
|---|---|---|---|
| B | Berol | P | Princ |
| BMA | BM Add | U | Upps |
| E | EMML | V | Vat |
| ME | Marḥa E. | Vi | Vindob. |

Bold face numbers, **1** to **72**, refer to texts as listed in the Concordance (p. 9) and transcribed in Chapter III (p. 69—147). None of the tabulations given here appear in the manuscripts in the same form. There we find only continuous sentences, e.g.:

> From our father Abraham to the Exodus of the Children (of Israel) from Egypt: 432 (years); and from the Exodus of the Children of Israel to the reign of David: 606 (years); and from David to the destruction of Jerusalem: 488 (years); etc.[1]

In my tabulations I allow myself abbreviations, like "Tower" for "Construction of the Tower", etc. On the other hand, I may add "years" where mere numbers are given in the text.

If in a tabulation columns are separated by a bold face vertical line, then everything to its left represents data found in the text, whereas everything to the right is my addition, e.g. the summation of years. Totals denoted by Σ at the end of a table are always my addition.

If, in my notes, I describe tables as "duplicates" or as "parallels", I disregard small variations, e.g. in spelling of names.

---

[1] From **4:** *B 84* $8^b$ I, 1—5. Other examples: Weld Blundell, Roy. Chron. p. 496f. (from *BM 815*) or Chaine, Chron. p. 112f. (from *BN 64*).

What makes two versions "duplicates" is agreement in the sequence and values of the numerical data. "Parallels" deal with essentially the same chronological period but may omit or add some events within the same sequence.

The "years" of the different eras are always "Coptic" (i.e., Alexandrian) years, beginning with Thoth (~ September). The eras commonly in use are denoted as follows:

| | |
|---|---|
| W | World, or "from Adam" (W 1) |
| J | Incarnation (J 0 = W 5500 = A.D. 7/8) |
| D | Diocletian (D 0 = W 5776) |
| G | "Grace", or "Mercy" (G 0 = W 5852 = D 76). |

For details cf. EAC p. 117 – Finally C = 532 years is the luni-solar cycle[2] that begins with W 1, W 533, etc.

In transcribing numbers I give, of course, what I consider the most plausible reading. It must be remembered, however, that 1 and 4, 6 and 7 could have been already misread, or miscopied, by the old scribes as easily as today, and that even a careful writing is no proof of the correctness of a number.

Discrepancies by $\pm 1^{y}$ may also be caused by the inconsistent use of cardinal or ordinal numbers of regnal years or eras. Years A.D. or B.C., I added only occasionally for orientation. Note that A.D. − n = B.C. n + 1.

[2] Cf. below p. 27.

## 2. Concordance

| | | | | |
|---|---|---|---|---|
| **1** | *B 84* | $7^b$ II, 24—$8^a$ I, 21 | | Flood→era G, Gabra Masqal |
| **2** | | $8^a$ I, 22—II, 7 | | Christ→Nicaea→Ṣaḥam 28 |
| **3** | | $8^a$ II, 7—24 | | Noah→Christ→Diocletian |
| **4** | | $8^a$ II, 24—$8^b$ I, 12 | | Flood→Diocletian |
| **5** | | $18^b$ I, 1—II, 3 | | $532^y$-cycles 1 to 13 |
| **6** | | $20^b$, 19—$21^a$ I, 28 | | Flood→Islam |
| **7** | | $22^b$ I, 2—II, 16 | cf. **55** | Flood→Gabra Masqal |
| **8** | *BM 754* | $4^a$ I, 1—9 | cf. **56** | Christ→Islam |
| **9** | *BM 815* | $17^a$ I, 15—$17^b$ I, 3 | | Noah→Christ, cycles |
| **10** | | $17^b$ I, 3—19 | | Christ |
| **11** | | $17^b$ I, 20—II, 17 | | Augustus→Vespasian; Gospels |
| **12** | | $17^b$ II, 17—$18^b$ I, 11 | cf. **57** 1 | Christ→Councils→Islam |
| **13** 1 | *BM 827* | $120^a$ I, 16—$120^b$ I, 9 | | Noah→Christ, cycles |
| **13** 2 | | $120^b$ I, 9—II, 3 | | Christ |
| **13** 3 | | $120^b$ II, 4—$121^a$ I, 3 | | Augustus→Vespasian; Gospels |
| **13** 4 | | $121^a$ I, 3—$121^b$ II, 1 | cf. **57** 2 | Christ→Councils→Islam |
| **14** | *BMA 16217* | $13^a$ I, 3—13 | | Augustus→Vespasian |
| **15** | | $19^a$ I, 2—$19^b$ I, 4 | cf. **58** | Noah→Islam, Vekuno Amlāk |
| **16** | | $20^a$ I, 4—II, 11 | | Adam→Abraham 175 |
| **17** | | $20^a$ II, 11—$21^a$ I, 12 | | Judges |
| **18** | | $21^a$ I, 12—II, 18 | | Kings of Judah |
| **19** 1 | *BMA 24995* | $30^b$ II, 1—$31^a$ I, 4 | cf. **60** | Noah→Gabra Masqal |
| **19** 2 | | $32^a$ I, 9—25 | | Flood→Christ→Gabra Masqal, cycles |
| **20** | *BN 160 A* | $2^a$, 1—11 | | the Persians |
| **21** | *C* | $2^a$, 20—28 | | the Ptolemies |
| **22** | *D* | $2^a$, 28—40 | | the Romans |

| | | | | |
|---|---|---|---|---|
| **23** | | $16^{a}$ I, 13—II, 18; $80^{b}$ I, 17—$81^{a}$ I, 5 | | Flood→Christ |
| **24** | | $76^{a}$ I, 1—13; I, 14—$76^{b}$ II, 8 | | Flood→Exodus; Judges |
| **25** | | $76^{b}$ II, 8—$77^{a}$ II, 5 | | Kings of Judah |
| **26** | | $77^{a}$ II, 5—$77^{b}$ I, 19 l. marg. | | Kings of Israel, etc. |
| **27** | | $77^{b}$ II, 1—$78^{b}$ II, 2 | | Adam→Abraham |
| **28** | | $78^{b}$ II, 2—$79^{a}$ II, 11 | | Judges |
| **29** | | $79^{a}$ II, 12—$80^{a}$ I, 9 | | Kings, captivity |
| **30** | *E 215* | $64^{a}$ II, 23—$64^{b}$ I, 28 | | $532^{y}$-cycles 1 to 13 |
| **31** | | $72^{b}$ I, 21—$73^{a}$ II, 7 | | Noah→Christ→Gabra M.→Yek. A., cycles |
| **32** | | $73^{a}$ II, 15—$73^{b}$ II, 8 | cf. **67** | same as **31** but no cycles |
| **33** | | $73^{b}$ II, 27—$74^{a}$ I, 23 | | Noah→Yekuno Amlāk |
| **34** | *E 2063* | $26^{a}$ II, 11—$27^{a}$ I, 8 | | Flood→Gabra Masqal, cycles |
| **35** | | $27^{a}$ I, 9—II, 16 | | Christ→Nicaea→Ṣaḥam 28 |
| **36** | | $27^{b}$ I, 1—II, 15 | | Noah→Christ→Diocletian |
| **37** | | $27^{b}$ II, 16—$28^{b}$ II, 12 | | Flood→Diocletian |
| **38** | | $44^{a}$ II, 5—$44^{b}$ II, 16 | | $532^{y}$-cycles 1 to 13 |
| **39** | | $46^{a}$ II, 7—$47^{a}$ II, 2 | | Flood→Islam |
| **40** | | $47^{a}$ II, 2—$47^{b}$ I, 8 | cf. **68** | Noah→Islam, Yekuno Amlāk |
| **41** | *E 2077* | $154^{b}$ I, 17—$155^{a}$ I, 6 | | Adam→Abraham, lives; to Exodus |
| **42** | | $155^{a}$ I, 7—23 | | Judges |
| **43** | | $155^{a}$ I, 23—II, 13 | cf. **69** | Kings; to Ezra |
| **44** | *ME* | p. 379 I, 23—380 I, 18 | | Flood, Exodus, Temple→Christ |
| **45** | *P 5884* | $7^{b}$ III, 10—20 | | Augustus→Vespasian, Temple |
| **46** | | $12^{b}$ II, 5—$13^{a}$ I, 14 | cf. **70** | Noah→Christ→Yekuno Amlāk |
| **47** | *U 3* | $63^{a}$, 2—$63^{b}$, 3 | cf. **71** | Noah→Christ, cycles |
| **48** | *V 1* | $205^{a}$ II, 41—$205^{b}$ II, 11 | | $532^{y}$-cycles 1 to 13 |
| **49** | | $205^{b}$ II, 28—$206^{a}$ I, 20 | | removal of Enoch→Abraham at 75, birth of Lēwi |

| | | | | |
|---|---|---|---|---|
| **50** | | $207^a$ I, 17—II, 1 | | $532^y$-cycles 1 to 11 |
| **51** | | $207^b$ II, 25—39 | cf. **72** | Flood→Christ |
| **52** | | $207^b$ I, 16—36 | | Christ→Nicaea→Ḍeḥam 28 |
| **53** | | $204^a$ II, 39—$205^a$ I, 8 | | Lēwi→Christ→D 76 |
| **54** | *Vi 6* | $2^b$, 17—26 | | Flood→Christ→Islam→Takla Hāymānot |
| **55** | *B 84* | $22^b$ II, 12—$23^a$ I, 26 | cf. **7** | Gabra Masqal→Lebna Dengel |
| **56** | *BM 754* | $4^a$ I, 9—II, 11 | cf. **8** | Yekuno Amlāk→Yāʿeqob, Zadengel |
| **57**1 | *BM 815* | $18^b$ I, 11—$19^a$ I, 14 | cf. **12** | Islam, Yekuno Amlāk→Iyāsu |
| **57**2 | *BM 827* | $121^b$ II, 1—$122^a$ II, 11 | cf. **13**4 | Islam, Yekuno Amlāk→Takla Hāymānot |
| **58** | *BMA 16217* | $19^b$ I, 2—$20^a$ I, 4 | cf. **15** | Islam, Yekuno Amlāk→Iyoʿās |
| **59** | | $21^b$ I, 1—17 | | Kings of Aksum (?) |
| **60** | *BMA 24995* | $31^a$ I, 1—II, 17 | cf. **19**1 | Nicaea, Gabra Masqal→Susenyos, Fasiladas |
| **61** | *BN 160 A* | $7^b$, 1—20; $90^a$ II, 6—14 | | Kings of Aksum |
| **62** | *B, C* | $7^b$, 20—25; 25—30 | | Christ→Conversion; Conversion→Qēsṭantinos |
| **63** | *D, E* | $7^b$, 30—36; $7^b$, 36—$8^a$, 21 | | Kings of Aksum→Conversion; Christ →Conversion |
| **64** | *F* | $8^a$, 21—28 | | Conversion→Delnaʿod |
| **65** | *G* | $8^a$, 29—36 | | Yekuno Amlāk→Yesḥaq |
| **66** | | $16^a$ II, 19—$16^b$ II, 9; $80^a$ I, 9—$80^b$ I, 16 | | Bāzēn→Lebna Dengel |
| **67** | *E 215* | $73^b$ II, 3—27 | cf. **32** | Gabra Masqal→Zareʿā Yāʿeqob |
| **68** | *E 2063* | $47^b$ I, 7—II, 7 | cf. **40** | Yekuno Amlāk→Yāʿeqob |
| **69** | *E 2077* | $155^a$ II, 21—III, 31 | cf. **43** | Zague→Takla Hāymānot |
| **70** | *P 5884* | $13^a$ I, 10—III, 20 | cf. **46** | Yekuno Amlāk→Yoḥanes |
| **71** | *U 3* | $63^b$, 3—$64^a$, 16 | cf. **47** | Christ→Zareʿe Yāʿeqob, cycles |
| **72** | *V 1* | $207^a$ II, 39—$207^b$ I, 9 | cf. **51** | Abrehā→Ḍeḥam |

## 3. Summary

All dates given in the following are based on the era W, using Coptic (i. e. Alexandrian) years. For other eras cf. p. 8.

### *A. Dates in Chronological Order*

| | | |
|---|---|---|
| 1455 | Enoch | **49** |
| 1569 | Noah | **27** |
| 1642 | Noah | **3, 36** |
| 1656 | Noah, birth | **9, 13** 1, **15, 19** 1, **31, 32, 40, 41, 46** |
| 1657 | Noah | **5, 30, 38** |
| *2000* | Noah | **47** |
| 2032 | Noah | **33** |
| 2068 | Flood | **51** |
| 2108 | Flood | **19** 2 |
| 2128 | Flood | **1, 34** |
| 2150 | Flood | **23** |
| 2242 | Flood | **3, 36, 49** |
| 2256 | Flood | **4, 6, 7, 9, 13** 1, **15, 19** 1, **31, 32, 37, 39, 40, 44, 46, 54** |
| 2257 | Flood | **24** |
| 2592 | Tower | **3, 36** |
| 2668 | Tower | **1, 34** |
| 2800 | Tower | **3, 6, 36, 39, 49** |
| 2827 | Tower | **7, 9, 13** 1, **15, 19** 1, **31, 32, 40, 46** |
| 2834 | Tower | **4, 37** |
| *3000* | Nimrud | **16** |
| 3040 | Abraham, Lord | **3, 36** |
| 3300 | Abraham, birth | **27** |
| 3324 | Abraham, 75 | **51** |
| 3328 | Abraham, birth | **7, 9, 13** 1, **15, 19** 1, **31, 32, 40, 46** |
| 3329 | Abraham, birth | **24** |
| 3349 | Abraham, left Chaldea | **23** |
| 3360 | Abraham, left Chaldea | **27** |
| 3388 | Abraham, 75 | **49** |

| | | |
|---|---|---|
| 3403 | Abraham, Lord | **4, 37** |
| 3413 | Isaac | **49** |
| 3432 | Abraham | **41** |
| 3440 | Abraham, Lord | **6, 39** |
| 3473 | Jacob | **49** |
| 3556 | Lewi | **53** |
| 3600 | Isaac | **47** |
| 3636 | Isaac | **33** |
| 3736 | Moses, birth | **53** |
| 3739 | Abraham, 75 | **1, 34** |
| 3753 | Moses, Exodus | **7, 9, 13** 1, **15, 19** 1, **31, 32, 40, 46** |
| 3754 | Exodus | **51** |
| 3779 | Exodus | **23** |
| 3790 | Exodus | **27** |
| 3816 | Exodus, Moses 80 | **53** |
| 3829 | Judges | **42** |
| 3835 | Exodus | **4, 37** |
| 3836 | Exodus | **24** |
| 3844 | Exodus | **44, 54** |
| 3856 | end of time in the desert | **53** |
| 3880 | Exodus, Moses | **3, 6, 36, 39** |
| 3958 | Judge Zatune'ēl | **53** |
| *4000* | Nāʿod 26 | **17** |
| 4030 | Moses | **47** |
| 4066 | Moses | **33** |
| 4152 | Exodus | **41** |
| 4161 | Judge Gēdeyon | **53** |
| 4169 | Exodus | **1, 34** |
| 4194 | Temple, construction | **51** |
| 4205 | Samuel | **28** |
| 4230 | Judge Yoftāhēl 6 | **53** |
| 4284 | end of Judges | **23** |
| 4315 | Judge Samson 20 | **53** |
| 4383 | Saul | **54** |
| 4395 | Samuel 20 | **53** |
| 4441 | David | **4, 37** |
| 4445 | Captivity | **43** |
| 4447 | David | **7, 9, 13** 1, **15, 19** 1, **31, 32, 40** |

| | | |
|---|---|---|
| 4456 | David | **46** |
| 4460 | Jerusalem, construction | **47** |
| 4495 | David, death | **53** |
| 4496 | Jerusalem, construction | **33** |
| 4609 | Temple, construction | **1, 34** |
| 4745 | Olympiad 1, 1 | **53** |
| 4800 | fall of Jerusalem | **23** |
| 4870 | return from Captivity | **23** |
| 4845 | Captivity | **54** |
| 4916 | Nebuchadnezzar | **7, 9, 13** 1, **15, 19** 1, **31, 32, 40, 46** |
| 4929 | destruction of Jerusalem | **4, 37** |
| 4930 | Captivity | **1** |
| 4933 | Captivity | **34** |
| 4965 | return from Captivity | **54** |
| 4973 | Alexander | **3, 36** |
| 4982 | construction of Temple | **44** |
| *5000* | Ezra | **1, 12, 33, 43, 47** |
| 5003 | Ezra | **34** |
| 5042 | Temple, Ezra | **44** |
| 5067 | Temple, Artaxerxes | **44** |
| 5128 | Captivity | **51** |
| 5171 | Alexander | **54** |
| 5173 | Alexander, death | **53** |
| 5181 | Alexander | **7, 9, 13** 1, **15, 19** 1, **31, 32, 40, 46** |
| 5182 | Alexander | **44** |
| 5192 | Alexander | **4, 37** |
| 5267 | Cleopatra | **3, 36** |
| 5457 | Cleopatra, death | **53** |
| 5471 | Cleopatra | **4, 37** |
| 5484 | Augustus | **44** |
| 5488 | Christ, birth | **54** |
| 5500 | Christ, annunciation . . . | **3, 36, 44** |
| 5500 | Christ | **6, 7, 9, 13** 1, **15, 19** 1, **22, 29, 31, 32, 39, 40, 46, 47, 53** |
| 5501 | Christ, birth | **4, 37, 44** |
| 5531 | Christ, baptism | **13** 2, **44** |
| 5533 | Christ, Crucif., Resurr. | **53** |

| | | |
|---|---|---|
| 5534 | Christ, Crucif., Resurr. | **10, 44** |
| 5536 | Christ, birth | **1, 2, 19** 2, **[33], [34], 35** |
| 5562 | Christ, birth | **51, 52** |
| 5569 | Christ, ascension | **2, 19** 2, **35** |
| 5574 | destruction of Temple | **11, 13** 3 |
| 5577 | destruction of Temple | **14, 45** |
| 5594 | Christ, ascension | **52** |
| 5745 | conversion of Ethiopia | **7, 8, 19** 1, **31, 32, 33, 71** |
| 5776 | Diocletian 1 | **3, 6, 7, 8, 12, 15, 31, 32, 36, 39, 40, 46, 53** |
| 5777 | Diocletian 1 | **4, 5, 30, 37, 38** |
| 5781 | conversion of Ethiopia | **33** |
| 5817 | Council of Nicaea | **8, 15, 40, 46** |
| 5835 | Council of Nicaea | **6, 7, 12, 13** 4, **19** 1, **31, 32, 39** |
| 5849 | G 1 | **34** |
| 5852 | G 1 = 11 C | **1, 2, 35, 53** |
| 5860 | Diocletian 1 | **52** |
| 5873 | Council of Constantinople | **8, 15, 40, 46** |
| 5893 | Council of Constantinople | **6, 7, 12, 13** 4, **39** |
| 5923 | Council of Ephesus | **8, 15, 40, 46** |
| 5929 | Gabra Masqal | **7, 19** 1, **31, 32, 71** cf. 6384 |
| 5944 | Council of Chalcedon | **8, 15, 40, 46** |
| 5948 | Council of Ephesus | **6, 12, 13** 4, **39** |
| 5963 | Gabra Masqal; Christian. | **33; 1, 34** |
| 5969 | Council of Chalcedon | **6, 12, 13** 4, 39 |
| 6035 | Abreha 1 | **72** |
| 6102 | Islam | **54** |
| 6109 | Islam | **6, 39** |
| 6114 | Islam | **8, 15, 40, 46** |
| 6139 | Islam | **12, 13** 4 |
| 6173 | Zague | **55, 67, 71** cf. 6629 |
| 6177 | Ṣaḥam 28 | **2, 35, 52, 72** |
| 6203 | Zague | **33** |
| 6300 | Yekuno Amlāk | **31, 33** |

| | | |
|---|---|---|
| 6306 | Yekuno Amlāk | **55, 67, 71** |
| | | cf. 6762 |
| 6336 | Yekuno Amlāk | **33** |
| 6380 | Sayfa Arʿād | **67, 71** |
| 6384 | Gabra Masqal = 12 C | **1, 19** 2, **34** |
| | | cf. 5929 |
| 6400 | Yesḥaq | **65** |
| 6450 | Yesḥaq | **55, 67** |
| | | cf. 6906 |
| 6471 | Zareʿa Yāʿeqob | **55, 67, 71** |
| | | cf. 6927 |
| 6544 | Lebna Dengel 1 | **55** |
| | | cf. 7000 |
| 6576 | Lebna Dengel 32 | **55** |
| | | cf. 7032 |
| 6629 | Zague | **65** |
| | | cf. 6173 |
| 6705 | Takla Hāymānot | **54** |
| 6762 | Yekuno Amlāk | **15, 40, 46, 56, 57** 1, **57** 2, **58, 70** |
| | | cf. 6306 |
| 6781 | Fasiladas | **60** |
| 6856 | Yesḥaq | **65** |
| | | cf. 6400 |
| 6906 | Yesḥaq | **56, 68, 70** |
| | | cf. 6450 |
| 6916 | 13 C | **1, 5, 34, 38, 48** |
| 6927/8 | Zareʿa Yāʿeqob | **56, 57** 1, **57** 2, **68, 70** |
| | | cf. 6471 |
| 6961/2 | Ba'eda Maryām | **56, 57** 1, **57** 2, **68, 70** |
| | | cf. 6505 |
| *7000/1* | Na'od; 13 C + 84 | **56, 57** 1, **57** 2, **58**; **60** |
| 7000/1 | Lebna Dengel | **68** |
| | | cf. 6544 |
| 7002 | Naʿod | **70** |
| 7138 | Fasiladas | **60** |
| 7171 | Takla Hāymānot | **58** |

| | | |
|---|---|---|
| 7201 | Takla Hāymānot | **57** 2 |
| 7390 | Yoḥanes | **70** |

### *B. Sequence of Events*

Here are collected important events for which multiple dates are found in our sources. For additional dates cf. the preceding section.

#### Noah

| | | |
|---|---|---|
| 1642 | **3, 36** | cf. Flood 2242 |
| 1656 | **9, 13** 1, **15, 19** 1, **31, 32, 40, 41, 46** | cf. Flood 2256 |
| 1657 | **5, 30, 38** | cf. Flood 2257 |
| 2000 | **47** | |
| 2032 | **33** | |

#### Flood

| | | |
|---|---|---|
| 2068 | **51** | |
| 2108 | **19** 2 | |
| 2128 | **1, 34** | |
| 2150 | **23** | |
| 2242 | **3, 36, 49** | cf. Noah 1642 |
| 2256 | **4, 6, 7, 9, 13** 1, **15, 19** 1, **31, 32, 37, 39, 40, 44, 46, 54** | cf. Noah 1656 |
| 2257 | **24** | cf. Noah 1657 |

#### Tower of Babel

| | | |
|---|---|---|
| 2592 | **3, 36** | |
| 2668 | **1, 34** | |
| 2800 | **3, 6, 36, 39, 49** | dispersion of languages |
| 2827 | **7, 9, 13** 1, **15, 19** 1, **31, 32, 40, 46** | |
| 2834 | **4, 37** | |

## Abraham

*Birth:*

| | |
|---|---|
| 3300 | **27** |
| 3328 | **7**, **9**, **13** 1, **15**, **19** 1, **31**, **32**, **40**, **46** |
| 3329 | **24** |

*Aged 75:*

| | |
|---|---|
| 3324 | **51** |
| 3388 | **49** |
| 3739 | **1**, **34** |

*Speaking with the Lord:*

| | |
|---|---|
| 3040 | **3**, **36** |
| 3403 | **4**, **37** |
| 3440 | **6**, **39** |

*Leaving Chaldea:*

| | |
|---|---|
| 3349 | **23** |
| 3360 | **27** |

## Moses, Exodus

*Birth:*

| | |
|---|---|
| 3736 | **53** |
| 3753 | **7**, **9**, **13** 1, **15**, **19** 1, **31**, **32**, **40**, **46** |

*Aged 80:*

| | | |
|---|---|---|
| 3816 | **53** | cf. 3736 |

*Exodus (and "Moses"):*

| | |
|---|---|
| 3754 | **51** |
| 3779 | **23** |
| 3790 | **27** |
| 3816 | **53** |
| 3835 | **4**, **37** |
| 3844 | **44**, **54** |
| 3880 | **3**, **6**, **36**, **39** |
| 4030 | **47** |

| | | |
|---|---|---|
| 4066 | **33** | |
| 4169 | **1, 34** | |

### David

| | | |
|---|---|---|
| 4447 | **7, 9, 13** 1, **15, 19** 1, **31, 32, 40** | |
| 4456 | **46** | |

*Death:*

| | | |
|---|---|---|
| 4495 | **53** | |

### Captivity

| | | |
|---|---|---|
| 4800 | **23** | |
| 4895 | **54** | |
| 4916 | **7, 9, 13** 1, **15, 19** 1, **31, 32, 40, 46** | |
| 4929 | **4, 37** | |
| 4930 | **1** | |
| 4933 | **34** | |
| 5128 | **51** | |

*Return after 70 years:*

| | | |
|---|---|---|
| 4870 | **23** | cf. 4800 |
| 4965 | **54** | cf. 4895 |

### Alexander

| | | |
|---|---|---|
| 4973 | **3, 36** | |
| 5171 | **54** | |
| 5181 | **7, 9, 13** 1, **15, 19** 1, **31, 40, 46** | = −311 A.D. |
| 5182 | **44** | |
| 5192 | **4, 37** | |

*Death:*

| | | |
|---|---|---|
| 5173 | **53** | |

### Cleopatra

| | | |
|---|---|---|
| 5267 | **3, 36** | |
| 5471 | **4, 37** | |
| 5493 | **19** 2 | |

*Death:*

| | | |
|---|---|---|
| 5457 | **53** | |

### Augustus

| | | |
|---|---|---|
| 5484 | **44** | |

### Christ

*Annunciation, Conception:*

| | | |
|---|---|---|
| 5500 | **3, 36, 44** | |

*Birth or Conception:*

| | | |
|---|---|---|
| 5488 | **54** | |
| 5500 | **6, 7, 9, 13** 1, **15, 19** 1, **22, 29, 31, 32, 39, 40, 46, 47, 53** | |
| 5501 | **4, 37, 44** | |
| 5536 | **1, 2, 19** 2, **33, 34, 35** | |
| 5562 | **51, 52** | |

*Baptism:*

| | | |
|---|---|---|
| 5531 | **13** 2, **44** | |

*Crucifixion, Resurrection:*

| | | |
|---|---|---|
| 5533 | **53** | |
| 5534 | **10, 44** | |

*Ascension:*

| | | |
|---|---|---|
| 5569 | **2, 19** 2, **35** | cf. 5536 |
| 5594 | **52** | |

### Destruction of the second Temple

| | | |
|---|---|---|
| 5574 | **11, 13** 3 | |
| 5577 | **14, 45** | |

### Conversion of Ethiopia

| | | |
|---|---|---|
| 5745 | **7, 8, 19** 1, **31, 32, 33, 71** | |
| 5781 | **33** | |

*Spread of Christianity:*

5963 **1, 34**

### Era Diocletian ("Martyrs")

5776 **3, 6, 7, 8, 12, 15, 31, 32, 36, 39, 40, 46, 53**

*from 532-year cycle:*

**5, 30, 38**
5777 **4, 37**

*actually era G:*

| | | |
|---|---|---|
| 5849 | **34** | |
| 5852 | **1, 2, 35, 53** | cf. 5776 |
| 5860 | **52** | |

### Church Councils

*Nicaea* (325 A. D.):

5817 **8, 15, 40, 46**
5835 **6, 7, 12, 13** 4, **19** 1, **31, 32, 39**

*Constantinople* (381):

5873 **8, 15, 40, 46**
5893 **6, 7, 12, 13** 4, **39**

*Ephesus* (431):

5923 **8, 15, 40, 46**
5948 **6, 12, 13** 4, **39**

*Chalcedon* (451):

5944 **8, 15, 40, 46**
5969 **6, 12, 13** 4, **39**

### Islam, era Hijra

| | | |
|---|---|---|
| 6102 | **54** | |
| 6109 | **6, 39** | |
| 6114 | **8, 15, 40, 46** | = 622 A. D. |
| 6139 | **12, 13** 4 | |

### Ṣaḥam

*year 28:*

| | | |
|---|---|---|
| 6177 | **2, 35, 52, 72** | |

### Gabra Masqal

| | | |
|---|---|---|
| 5929 | **7, 19** 1, **31, 32, 71** | cf. 6384 |
| 5963 | **33** | |
| 6384 | **1, 19** 2, **34** | cf. 5929 |

### Yekuno Amlāk

| | | |
|---|---|---|
| 6300 | **31, 33** | |
| 6306 | **55, 67, 71** | cf. 6762 |
| 6336 | **33** | |
| 6762 | **15, 40, 46, 56, 57** 1, **57** 2, **58, 70** | cf. 6306 |

### Takla Hāymānot

| | |
|---|---|
| 6693 | **54** |
| 7170 | **58** |
| 7201 | **57** 2 |

### Naʿod

| | |
|---|---|
| 7000 | **56, 57** 1, **57** 2 |
| 7001 | **58** |
| 7002 | **70** |

## *C. Main Topics and Data*

| | |
|---|---|
| 532-year Cycle: | **5, 9, 30, 31, 32, 38, 48, 50** |
| Patriarchs: | **16, 27, 41** |
| Judges: | **17, 24, 28, 42, 53** |
| Kings of Judah: | **18, 25, 29, 43** |
| Kings of Israel, Edom: | **26** |
| Persians: | **20** |
| Ptolemies: | **21** |
| Romans: | **11, 13** 3, **14, 22, 45** |
| Christ: | **13** 2, **44** |

Ethiopic Kings: **55** to **72**

Basic Dates:

| | | | |
|---|---|---|---|
| Noah, birth | 1656 | Nebuchadnezzar | 4916 |
| Flood | 2256 | Ezra | 5000 |
| Tower | 2827 | Alexander | 5181 |
| Abraham, birth | 3328 | Christ | 5500, 5536 |
| Exodus | 3753 | Diocletian 0 | 5776 |
| David | 4447 | Islam | 6114 |
| | | Yekuno Amlāk | 6306, 6762 |

## 4. Bibliographical Abbreviations

AVI-YONAH, Michael: Geschichte der Juden im Zeitalter des Talmud, in den Tagen von Rom und Byzanz. Studia Judaica 2, Berlin 1962.

BEZOLD, Carl: Kebra Nagast. Bayer. Akad. d. Wiss., I Kl., 23, 1. München 1905.

BIBLE: Names and dates are taken from The New Oxford Annotated Bible with the Apocrypha. Oxford Univ. Press, 1977.

BICKERMAN, Elias: Four Strange Books in the Bible. New York 1968.

CHAINE, M.: La chronologie des temps chrétiens de l'Égypte et de l'Éthiopie. Paris 1925.

CHI: The Cambridge History of Iran. Cambr. Univ. Press.

CONTI ROSSINI, Ch. [1909]: Les listes des rois d'Aksoum. J. As. 14 (1909), p. 263—320.

DILLMANN, A.: Catalogus codicum manuscriptorum qui in Museo Britannico asservantur. Pars III. Codices Aethiopicos amplectens. London 1847.

DILLMANN, A.: Catalogus codicum manuscriptorum Bibliothecae Bodleianae Oxoniensis, VII. Codices Aethiopici. Oxford 1848.

DILLMANN, A. [1853]: Zur Geschichte des abyssinischen Reichs. ZDMG 7 (1853), p. 338—364.

EAC: See Neugebauer, EAC.

GARSTANG, John: Joshua, Judges, London 1931.

GELZER, Heinrich: Sextus Julius Africanus und die Byzantinische

Chronographie. I, Leipzig 1880, II, 1898 (reprint Hildesheim 1978).

GRAF, Georg: GAL II: Geschichte der christlichen arabischen Literatur II, Studi e Testi 133. Città del Vaticano, 1947.

HAKLUYT SOC.: Ser. II No. 107: Some Records of Ethiopia 1593—1646 (London 1959).

HAMMERSCHMIDT, Ernst: Äthiopische Kalendertafeln, Wiesbaden, Steiner, 1977.

J. AS.: Journal Asiatique.

KRUSCH, Bruno: Studien zur christlich mittelalterlichen Chronologie. Leipzig 1880.

MAURO DA LEONESSA [1943]: Un trattato sul calendario redatto al tempo di re ʿAmda-Ṣyon I. Rassegna di Studi Etiopici (1943), p. 302—326.

NEUGEBAUER O.: EAC: Ethiopic Astronomy and Computus. Österr. Akad. d. Wiss., Phil.-Hist. Kl., Sitzungsberichte 347 (1979).

NEUGEBAUER, O. [1981]: On the Spanish Era. Chiron 11 (1981), p. 371—380.

NEUGEBAUER, O. [1985]: The Chronological System of Abu Shaker (A. H. 654). Annals of the New York Academy of Sciences. Vol. 500 (1987), pp. 279—293.

NEUGEBAUER, O. [1987]: Byzantine Chronography, a Critical Note. Byzantinische Zeitschrift 80 (1987), pp. 330—333.

NEUGEBAUER, O. [1988]: Abu Shaker's "Chronography". Österr. Akad. d. Wiss., Phil.-Hist. Kl., Sitzungsberichte 498 (1988).

OPPENHEIM, Leo: Ancient Mesopotamia, rev. ed., Chicago Univ. Press, 1977.

PERRUCHON, J.: Vie de Lalibala. Paris, 1892.

PG: Patrologia Graeca.

PRITCHARD, James B. (ed.): Solomon & Sheba, London, Phaidon, 1974.

RÜHL, Franz: Chronologie des Mittelalters und der Neuzeit, Berlin 1897.

SACHS, A. [1977]: Achaemenid Royal Names in Babylonian Astronomical Texts. Am. J. of Ancient History 2 (1977), p. 127—147.

SKEAT, T. C.: The Reigns of the Ptolemies. Münchener Beiträge zur Papyrusforschung 39. München 1954.

TAMRAT, Tadesse: Church and State in Ethiopia 1270—1527. Oxford Clarendon Press, 1972.

ULLENDORFF, Edward: The Ethiopians. Oxford Univ. Press, 1960.
WELD BLUNDELL, H.: The Royal Chronicle of Abyssinia 1769—1840, with Translation and Notes. Cambridge Univ. Press, 1922.
ZDMG: Zeitschrift der Deutschen Morgenländischen Gesellschaft.

# II. THE CHRONOLOGICAL MATERIAL

## 1. World History

### *A. From the Deluge to Diocletian (and beyond)*

The historical material in the texts discussed here is centered on some major events from Biblical history (e. g., Deluge or Exodus), the Persian Kings (return from Exile, Second Temple), Alexander and the Ptolemies, the Romans from Augustus to Diocletian and the Church-Councils. About 20 of these texts extend their chronology into Ethiopic history (cf. Section 4, p. 55).

Each one of these tabulations enumerates at most some 15 to 20 events. There is no evidence for any wider historical interest. Neither Egyptian nor Babylonian dynasties are ever mentioned (as is the case in Byzantine chronographic treatises). Islam is only represented by the date of the Hijra. Ethiopic "World History" remained on a very modest level.

### *B. The 532-year Cycle and Chronology*

The shortest cycle in which the dates of Easter repeat themselves is the 532-year cycle. This cycle contains the factor 19 because the lunar phases (e. g. full moons) coincide every 19th year with the same solar date. Return of the weekdays requires intervals of 7 years (because $365 \equiv 1 \bmod 7$); finally 4 years take care of the Alexandrian ("julian") intercalation pattern. And $19 \cdot 7 \cdot 4 = 532$.

Obviously it was the Passover- and Easter-computus that kept the 532-year cycle in constant use for many centuries. But this cycle plays an unexpectedly important role in different aspects of Ethiopic chronology, not only in general, but also for our understanding of the Ethiopic kinglists, a topic which we shall discuss in detail in a later section (p. 55 ff.).

The significant role of the 532-year cycle for chronology is obvious from the fact that the era G (of "Grace" or "Mercy", meḥrat) has its starting point exactly at a distance of $11C = 5852^y$ from the first year of the era of the "World" (or "from Adam"). Nevertheless this is only an artificial construction of the theologians and Byzantine chronographers, actually based on the Roman era "Diocletian", later renamed by Christian authors "years of the Martyrs" (samâʿetât)[1]. The close parallelism between the eras D and G explains that repeatedly years of the era G are called years of the "Martyrs"[2]. Apparently the era G was invented for the Ethiopian church; Abu Shaker, writing in Cairo, uses only the era Diocletian.

In the present material the 532-year cycles of the era W or G are very useful in securing chronological fixed-points. For example the birth of Noah is placed in the year 60 of the 4th cycle[3] or 473 years before the beginning of the 5th cycle[4]. Consequently the year of Noah's birth ("Noah year 1") is

$$1596 + 60 = 2128 - 472 = 1656 \quad \text{where} \quad 1596 = 3 \cdot 532,\ 2128 = 4 \cdot 532.$$

Similarly the first year of the era Diocletian is described as

$$5777 = 5853 - 76 \quad \text{and} \quad 5853 = 11 \cdot 532 + 1$$

or[5]

$$5776 = 5320 + 456 \quad \text{i. e.} \quad 10C + 456$$

for the year $D = 0$.

Table 1 gives a list of the texts that correlate cycle years and historical dates[6]. They fall into two different groups: the first five texts refer (until Ptolemy Philadelphus) only to strictly biblical events, beginning with Adam, whereas the other three texts, starting with Noah, follow the more secular orientation of the majority of our tabulations.

---

[1] A complete 532-year Easter cycle for the era G is published in EAC p. 59 to 63, for the era W in Abu Shaker, p. 193 to 198.

[2] This error is found, e. g., in the texts **1, 2, 34, 35, 48.**

[3] In **31** and **32.**

[4] In **5, 30, 38.**

[5] In **31** and **32.**

[6] Isolated references to cycle years are found, e. g., in **12** for the Council of Constantinople ($5893 = 11C + 41$), or in **7** and **19** 1 for Christ and for Nicaea.

*Table 1*

| n | A = nC+1 | B = | 5, 30, 38 | A−B +1 | | **48** | **50** | | **9, 13 1, 31, 32** | n | nC |
|---|---|---|---|---|---|---|---|---|---|---|---|
| | 1 | Adam | 1 | | | 1 | 1 | | | | |
| 1 | 533 | Ēnos | 98 | 436 | | 98 | 98 | | | | |
| 2 | 1065 | Yārēd | 105 | 961 | | 105 | 105 | | | | |
| 3 | 1597 | Lāmēh | 123 | 1475 | | 143 | 113 | | | | |
| 4 | 2129 | Noah | 473 | 1657 | | 87 | 87 | birth of Noah | 1656 = 3 C + 60 | 3 | 1596 |
| 5 | 2661 | Ebēr | 129 | 2533 | | 23 | 23 | Flood | 2256 = 4 C + 128 | 4 | 2128 |
| 6 | 3193 | Nākor | 5 | 3189 | | 29 | 29 | Tower | 2827 = 5 C + 167 | 5 | 2660 |
| 7 | 3725 | Embarm | 43 | 3683 | Abraham | 65 | — | Abraham | 3328 = 6 C + 136 | 6 | 3192 |
| 8 | 4257 | Gēdēwon | 2 | 4256 | Agālom[1] | 7 | 36 | Moses | 3753 = 7 C + 29 | 7 | 3724 |
| 9 | 4789 | Ḥezeqyās | 23½ | 4766½ | Yonātān | 8 | 38 | David | 4447 = 8 C + 191 | 8 | 4256 |
| 10 | 5321 | Ptolemy | 16 | 5306 | Ptol. Philad. | 37 | 38 | Nebukadn. | 4916 = 9 C + 128 | 9 | 4788 |
| 11 | 5853 | Martyrs | 77 | 5777 | Diocletian[2] | 2 | — | Alexander | 5181 + 393 | | |
| 12 | 6385 | Martyrs | 608 | | Diocletian | 534 | | Christ | 5500 = 10 C + 180[3] | 10 | 5320 |
| 13 | 6917 | total: | 6916 | | | | | Conversion | 5745 + 425[4] | | |
| | | | | | | | | Diocletian | 5776 + 456 | | |
| | | | | | | | | Nicaea | 5835 + 515[5] | | |
| | | | | | | | | Gabra Masqal | 5929 = 11 C + 77[6] | 11 | 5852 |

[1] Cf. Judges (Table 4 p. 35)
[2] error for era G
[3] same in **7**, **19** 1, **47**
[4] same in **47**
[5] same in **7**, in the form 11 C—17
[6] same in **19** 1, and **71**; actually $456^y$ later, i. e. 6384 = 12 C

References to cycle-years are frequently found in chronological lists. A collection of such cases is given in Table 21 (p. 134) which illustrates the widespread use of the 532-year cycle numbers.

As we have shown above, these two groups agree in two fundamental dates: birth of Noah (1656) and Era Diocletian (5776). The other dates in the second group (**9**, **13** 1, **31** and **32**[7]) agree with the majority of dates for Noah to the Council of Nicaea found in other texts[8]. The first group contains several variants, the origin of which I cannot explain, excepting the entries under "Diocletian" which are actually years of the era "Grace".

Note: The 532-year cycles occur also in connection with calendrical parameters: *Vat 1* 203[a], 16—204[a] II, 5 discusses the determination of the parameters[9] i, e, ep, t for the first years of the 532-year cycles 1, 3, 6, 9, 11, 13, naming the last three cycles "years of the era Diocletian", again misinterpreting years of the era G.

This is not the place to discuss the development of the 532-year cycle into the era of the World in which the year of the Creation coincides with the first year of this cycle. It suffices to note that Panodorus (around A. D. 400) and Annianus (412) seem to be the inventors of this chronological system[10] in which the birth of Christ is associated with the year 5500. For the year 5536 and its conjectural connection with the "Spanish Era" cf. below p. 52.

Gelzer quotes approvingly[11] a statement by Unger saying that Annianus was "weiter nichts als ein handwerksmäßiger Passacalculator" (an otherwise unknown profession). Obviously neither Unger nor Gelzer realized the significance of the 532-year cycle, and hence of the Easter computus, for the connection between biblical chronology and the era Diocletian, a connection which prepared the road to the era "A. D.".

---

[7] For a close parallel to **32** cf. Chaine, Chron., p. 112f. (*BN 64*, 58[b]—60[b]).

[8] Example: from the left group we find for Embarm 1: 3725 − 42 = 3683, from the right group Moses 1 = 3753; hence Moses 1 = Embarm 70 in agreement with the age given to Moses' father at the time of his birth (cf., e. g., Gelzer, Afric. I p. 86).

[9] Cf. EAC p. 11.

[10] Cf. Gelzer, Afric. II p. 191; also Rühl, Chron. p. 117.

[11] Afric. II p. 190.

## 2. Biblical History

The chronological lists dealing with Biblical history are considerably more extensive within each of their sections (Patriarchs, Judges, Kings) than the lists concerning "World History". This fact is easy to explain: the names listed here follow strictly the Biblical narrative and thus constitute a detailed chronological survey of extensive historical periods.

### *A. The Patriarchs*

We have three major texts, **41**, **27**, **16**, which, however, not only differ in the form in which they present their data but also deviate in many details. The subsequent tabulation (Table 2) condenses these lists to their basic numerical contents. Hence it is necessary to remark briefly on the actual form of presentation.

No. **41** follows, from Adam to Noah, the pattern:

*A* generated *B* at the age (a) and lived thereafter (m) years, hence reached a lifetime of (n) [ = a + m] years.

After Noah, the last information (m) is omitted.

No. **27** says:

*A* had lived (a) years when *B* was born. *B* had lived (b) years when *C* (in the year a + b) was born, etc.

In this way the birth of Abraham is given the round date 3300.

No. **16** is in bad shape. The text gives only the years of life from Adam to Sêruḥ and Nâmrud but on the margin are written corrections for almost all dates as well as years for Nâkor to Abraham.

The existence of many discrepancies in these lists is not surprising since already the Biblical text survived in two versions, the Septuagint and the Masoretic version (cf. Table 3). The Ethiopic texts follow, as is to be expected, usually the LXX but **41** has two Masoretic data (1. 8 and 9) and **16** gives (1. 9) 776, probably a copyist error for 777 of the Masoretic version. The total

*Table 2*

| | 41 father at | 41 lived thereafter | 41 hence total | 27 | 27 father at | 27 total | 16 lived text | 16 lived margin | |
|---|---|---|---|---|---|---|---|---|---|
| Adam | 230 | 700 | 930 | | 230 | | 950 | 930 | |
| Sēt | 205 | 707 | 972 | | 205 | 430 | 917 | 902 | |
| Hēnos | 190 | 715 | 905 | | 190 | 620 | | 905 | |
| Qāynān | 170 | 740 | 910 | | [100] | 725 | 920 | 910 | |
| Mālāleʿel | 165 | 640 | 805 | | 165 | 890 | 805 | 1095 | |
| Yārēd | 162 | 800 | 962 | | 162 | 1052 | 972 | 962 | |
| Hēnoḫ | 165 | 200 | 365 | | 162 | 1214 | | | |
| Mātusalū | 187 | 782 | 969 | | 167 | 1384 | 999 | | |
| Lameḫ | 182 | 595 | 777 | | 128 | 1569 | 776 | 746 | |
| Adam→Noah | 1656 | | | | | | | | |
| Noah | 500 | (350) | 950 | | 500 | 2069 | 950 | | |
| Sēm | 100 | 500 | | | 100 | 2169 | 700 | 500 | |
| Arfāskad | 135 | 335 | | | 135 | 2304 | 465 | 440 | |
| Qāynān | 130 | 440 | | | 130 | 2434 | 430 | 440 | |
| Sālā | 130 | 430 | | | 130 | 2564 | | 430 | |
| Ēber | 134 | 430 | | | 130 | 2694 | 434 | | |
| Fālēq | 130 | 270 | | | 130 | 2824 | 430 | | |
| Rāgew | 232 | 269 | | | 132 | 2956 | 230 | 230 | |
| Sēroh | 136 | 200 | | | 135 | 3091 | 236 | 206 | |
| Nākor | 79 | 129 | | | 109 | 3200 | (69) | 129 | Nākor |
| Tārā | 70 | 250 | | | 100 | 3300 | | 205 | Tārā |
| Noah→Abreh. | 1776 | | | Abreham | 60 | 3360 | | 127 | Sārā |
| Adam→Abreh. | 3432 | | | | | | | 175 | Abreh. |
| Abreh.→Yeshaq | 100 | 75 | | | | | | | |
| →Yaʿeqob | 60 | | | | | | | | |
| →Pharao | 130 | | | | | | | | |
| in Egypt | 430 | | | in Egypt | 430 | 3790 | | | |
| Abreh.→Exod. | 720 | | | | | | | | |
| Adam→Exodus | 4152 | | | | | | | | |

of 1642 years from Adam to the birth of Noah, however, is only twice represented in our material (**3** and **36**), while the Masoretic total 1656 is nine times attested and underlies also the commonly used date 2256 for the Deluge in Noah's year 600 (cf. above p. 17). Some of the marginal corrections in **16** come from LXX but the majority of the dates in **16**, text as well as margin, have no obvious source.

It is of interest to note that the discrepancies in the chronology of the patriarchal age, in particular about Gen. 11, 10—26, caused concern as early as in the 12th century. Abu Shaker in Ch. 55 of his "Chronography"[12] polemizes against "Targum 70 to the Torah", probably a commentary (not a "translation", as targum would mean in common Hebrew terminology). He calls this Targum "absurd" (beṭul) probably because it is based on the short Masoretic chronology (cf. Table 3 p. 34) which he quotes and which implies that Noah (who supposedly lived 950 years) could have seen not only the construction of the Tower and the dispersion of the languages (in the time of Eber) but even Abraham. Unfortunately Abu Shaker does not provide additional data which would allow us to reconstruct the details of the competing chronological systems.

## *B. The Judges*

We have four Ethiopic texts, **17, 24, 28, 42,** which give a list of the reigns of the Judges (and thereafter of the Kings of Judah): cf. Table 4. For variants found in the works of Byzantine chronographers, as collected by Gelzer, cf. my note [1987].

The first item, the 40 years of Wandering in the Desert under Moses' leadership, follows biblical tradition. Its version of $80^{y}$ for Naod and Sêmêgâr is represented in **24** and **42** whereas **17** and **28** give for Naod $80^{y}$ and for Sêmêgâr $25^{y}$ which mistakenly reflects a Byzantine splitting of the 80 years into 55+25.

The absolute date in **17**: Naod 26 = W 4000, leads to Naod 1 = W 3975. The preceding 141 years give us the time since the Exodus which thus would fall in W 3834[13]. The stay of 430

[12] Cf. for Abu Shaker: Neugebauer [1988], in particular p. 104.

[13] W 3835 in **4** and **37**; cf. also note 17.

*Table 3*

| | | Masoretic | | | | LXX | | | | Deviations from LXX | | | |
|---|---|---|---|---|---|---|---|---|---|---|---|---|---|
| | | father of | at age | lived thereafter | total | | at age | lived thereafter | total | **41** at | **27** at | **16** text | **16** margin |
| 1. | Adam | Seth | 130 | 800 | 930 | | 230 | 700 | 930 | | | 950 | |
| | Seth | Enosh | 105 | 807 | 912 | | 205 | 707 | 912 | | | 917 | 902 |
| | Enosh | Kenan | 90 | 815 | 905 | | 190 | 715 | 905 | | | | |
| | Kenan | Mahalalel | 70 | 840 | 910 | | 170 | 740 | 910 | | | 920 | |
| 5. | Mahalalel | Jared | 65 | 830 | 895 | | 165 | 730 | 895 | | | | 1095 |
| | Jared | Enosh | 162 | 800 | 962 | | 162 | 800 | 962 | | | 972 | |
| | Enoch | Methuselah | 65 | 300 | 365 | | 165 | 200 | 365 | | 162 | | |
| | Methuselah | Lamech | 187 | 782 | 969 | | 167 | 802 | 969 | 187 | | 999 | |
| | Lamech | | 182 | 595 | 777 | | 188 | 565 | 753 | 182 | | 776 | 746 |
| 10. | Adam → Noah | 1656 | | | | 1642 | | | | | | | |
| | Noah | Shem | 500 | | | | 500 | | | | | | |
| | Shem | Arpachshad | 100 | 500 | | | 100 | 500 | | | | 700 | |
| | Arpachshad | Shelah | 35 | 403 | | | 135 | 400 | | | | 465 | 440 |
| | | | | | | Kainan | 130 | 330 | | | | 430 | |
| 15. | Shelah | Eber | 30 | 403 | | | 130 | 330 | | | | | |
| | Eber | Peleg | 34 | 430 | | | 134 | 270 | | | | 434 | |
| | Peleg | Reu | 30 | 209 | | | 130 | 209 | | | | 430 | |
| | Reu | Serug | 32 | 207 | | | 132 | 209 | | 232 | | 230 | 230 |
| | Serug | Nahor | 30 | 200 | | | 130 | 207 | | 136 | 135 | 236 | 207 |
| 20. | Nahor | Terah | 29 | 119 | | | 179 | 200 | | 79 | 109 | | 129 |
| | Terah | Abraham | 70 | | d.: 205 | | 70 | | d.: 205 | | 100 | | 205 |
| | Sarah | | | | d.: 127 | | | | d.: 127 | | | | |
| | Abraham | | | | d.: 175 | | | | d.: 175 | | | | |

*Table 4*

| | 17 | 24 | 28 | | 42 | Bible | |
|---|---|---|---|---|---|---|---|
| subject to Pharaoh | 430 | | | [3790] | | Desert | 40 |
| Moses | 40 | 40 | 40 | | | Joshua | — |
| Faneḥas (or Iyasus) | 25 | 34 | 31 | 3861 | | | |
| Kursa | 8 | 8 | 8 | | 8 | Cushan-rishataim | 8 |
| Gotolyāl | 50 | 50 | 5 | | 50 | Othniel | 40 |
| Eglom, king of Moab | 18 | 18 | 18 | 3892 | 18 | Eglon | 18 |
| Nāʿod | 80 | 80 | 80 | | 80 | Ehud | 80 |
| Nāʿod 26 W 4000 | | | | | | | |
| Sēmēgēr | 25 | — | 25 | 3907 | — | Shamgar | — |
| Iyāmin | 25 | 25 | — | | — | Jabin | 20 |
| Diborā and Bārq | — | 40 | 40 | | 7 | Deborah and Barak | 40 |
| Midians | 7 | — | 7 | 3954 | 40 | Midianites | 7 |
| Gēdiwon | 44 | 40 | 40 | | 40 | Gideon | 40 |
| Abēmēlēk | 3 | 3 | 3 | | — | Abimelech | 3 |
| Tolā | 23 | 23 | 23 | 4020 | 23 | Tola | 23 |
| Iyāēr, the Gileadite | 22 | 22 | 22 | | 22 | Jair, the Gileadite | 22 |
| Philistines | 18 | — | 18 | | 18 | Philistines | 18 |
| Yoftāhē | 7 | [6] | 6 | 4066 | 6 | Jophthah | 6 |
| Ḥasēbon | 10 | 7 | 7 | | 7 | Ibzan | 7 |
| Sēlom | 10 | — | 10 | | 10 | Elon | 10 |
| Lābon | 8 | 8 | 8 | 4091 | 8 | Abdon | 8 |
| Philistines | 40 | 40 | 40 | | 40 | Philistines | 40 |
| Samson | 20 | 20 | 20 | 4151 | 20 | Samson | 20 |
| no ruler | 12 | — | 12 | | Exodus to | | |
| Ēli | 40 | 20 | 20 | | Judges: 397 | | |
| Sāmuēl | 22 | 20 | 22 | 4205 | Adam: 3829 | | |
| | Σ 557 | 504 | 505 | | | | Σ 450 |

years in Egypt then gives W 3404 for "Abraham", presumably the date for his conversation with the Lord[14], hence W 3330 for his birth[15]. Obviously 40 years should have been given to Debora and Barq. Combining this with the above-mentioned correction for Naod and Sêmêgâr, one obtains for the end of Samson W 4332 and W 4406 for Samuel.

The text **24** defines the Exodus in the following fashion:

| | |
|---|---|
| Adam → Flood | 2257[16] |
| Flood → birth of Abraham | 1072 |
| Abraham → Exodus | 507 |
| thus Exodus: | 3836 |

instead of 3834 in **17**. Probably both cases should be emended to W 3835[17].

Finally: correcting the omission of 7 years for the Midians (Table 4, l. 12) and of 18 years for the Philistines (l. 17) lead for the end of Samson to W 4323 (and hence for Samuel only 40 years more).

In **28** the 5 years for Gotolyar (l. 5) are a scribal error for 50. The 25 years for Sêmêgâr (l. 9) belong to the next entry but do not alter the total. In this way, we obtain 294 years until Samson and 54 years more until Samuel. Assuming again W 3836 for the Exodus gives 4332 for Samson and 4386 for Samuel (hence 4387 for Saul 1). This, however, does not agree with the years of the World, given in the text, for several reasons: first, as the date of the Exodus is assumed W 3790 (as it follows from the subsequent entries); secondly, the incorrect 5 years (Table 4, l. 5), instead of 50, are reckoned as 5 in the total; thirdly, 3892 + 105 are taken (in l. 9) as 3907 instead of 3997. This leads to the erroneous dates 4151 for Samson and 4205 for Samuel, both 181 years too early.

The numbers given in **42** make a total of 397 years and this number is also mentioned in the text as "from Exodus to the Judges". It follows, however, from the names listed that an interval of about 70 or 100 years is missing. The additional statement "and from Adam 3829 years" makes only sense as date of the Exodus

[14] Again **4** and **37**.

[15] W 3329 in **24**. The commonly used date is 3328 (cf. p. 18).

[16] Ordinarily 2256.

[17] Attested in **4** and **37** as well as in Eutychius (Said ibn Baṭriq, about A. D. 900), Gelzer, Afric. II p. 409.

(cf. above W 3834/6) which would give for Samson a date between 4300 and 4330. Obviously the dates near the end of the Judges were only very insecurely transmitted.

In **53**[18] one finds a short intrusion that concerns the time of the Judges and agrees well with the biblical chronology[19]:

*Table 5*

| **53** | | | Bible | |
|---|---|---|---|---|
| Exodus | | 3818 | | |
| in the Desert | 40$^{y}$ | 3856 | | |
| until Tusa year 26 | 57 | 3913 | Cushan-rishataim | 8$^{y}$ |
| until Zatuneēl | 45 | 3958 | Othniel | 40 |
| until Gēdewon | 203 | 4161 | Gideon | 205 |
| until Yeftāḥēl year 6 | 69 | 4230 | Jephthah | 72 |
| until Samson year 20 | 85 | 4315 | Samson | 85 |
| | | 402 | | |
| until Samuel year 20 | 8[0] | 4395 | | |
| until the death of David | [100] | 4495 | | |

All these manuscripts illustrate the influence of trivial scribal errors in generating what had been taken to be different systems of chronology but in fact is only the result of centuries of copying more or less faulty predecessors.

## *C. The Kings of Judah*

The texts **18, 25, 29, 43** (cf. Table 6)[20] are the direct continuation of the corresponding four texts that concern the time of the Judges (cf. the preceding section).

Text **18** ends with Amêzyâs. The totals of the remaining lists are correctly computed only in **29**. Difficult to explain is the fact

---

[18] Cf. below p. 42, Table 9.

[19] The 8[0] in l. 9 of Table 5 corrects a scribal error 85 (cf. l. 7). The total is correct.

[20] The spelling of names follows mainly **43**.

*Table 6*

| | 18 | 25 | 29 | | 43 | | Bible | |
|---|---|---|---|---|---|---|---|---|
| | | | | W 4205 | | | | |
| Sā'ol | 40 | 40 | 40 | | 'Ēli | 20 | [40] | Saul |
| Dāwit | 40 | 40 | 40 | | | 40 | 40 | David |
| Salomon | 40 | 40 | 40 | 4325 | | 40 | 40 | Solomon |
| Robe'ab | 17 | 17 | 17 | | | 17 | 17 | Rehoboam |
| 'Abiyu | $7^y69^d$ | $6^y8^d3^h$ | $6^y8^d[3]^h$ | | | 3 | 3 | Abiyam |
| 'Asā | 41 | 44 | 41 | 4389 | | 41 | 41 | Asa |
| 'Iyosāfeṭ | 25 | 25 | 25 | | | 25 | 25 | Jehoshaphat |
| 'Iyorām | 8 | 8 | 8 | | | 8 | 8 | Jehoram |
| 'Akāzyās | 1 | 10 | 1 | | | 4 | 1 | Ahaziah |
| Gotolyā | 8 | 10 | 6 | 4409 | | 7 | 7 | Atheliah |
| 'Iyo'as | 40 | 40 | 40 | | | 40 | 40 | Jehoash |
| 'Amēsyās | 30 | 29 | 29 | | | 29 | 29 | Amaziah |
| 'Azāryās | | 52 | 52 | 4550 | | 52 | 52 | Uzziah |
| 'Iyo'atām | | 16 | 16 | | | 16 | 16 | Jotham |
| 'Ākaz | | 17 | 16 | | | | 16 | Ahaz |
| Ḥezegyās | | 29 | 29 | | | 29 | 29 | Hezekiah |
| Menāsē | | 52 | 55 | | | 55 | 55 | Manasseh |
| 'Amoṣ | | $2^y12^d$ | $2^y12^d$ | | | 2 | 2 | Amon |
| Yoseyās | | 34 | 31 | 4699 | | 81 | 31 | Josiah |
| 'Iyo'akaz | | $3^m$ | $3^m$ | | | | $3^m$ | Jehoahaz |
| 'Ēliyāqim | | $14^y14^d$ | $11^y14^d$ | | | 11 | 11 | Jehoiakim |
| 'Iyo'aqim | | $3^m$ | $3^m$ | | | $3^m$ | $3^m$ | Jehoiachin |
| Mātān | | 14 | 11 | 4730 | Nātānyān | 21 | 11 | Mattaniah |
| total | | $526^y7^m3^d$ | $516^y7^m4^d3^h$ | | | $616^y3^m$ | | |
| Σ | | $539^y7^m4^d3^h$ | $516^y7^m4^d3^h$ | | | $541^y3^m$ | $514^y6^m$ | |

that some of the intervals count days and hours. Astrological reasons seem to me historically unlikely since Ethiopic material in general shows no interest in astrology.

The date W 4447 for David, mentioned in eight instances (cf. above p. 19) indicates that it is a fixed-point in Biblical history.

### *D. Kings of Israel and of Edom*

The only text in our material which concerns kingdoms contemporary with the kings of Judah is **26** from *BN 160.* Its main part (cf. Table 7) is a list of kings of Israel which agrees very well (excepting a few minor deviations) with the regnal years mentioned in Kings I and II. The Ethiopic text says that 18 kings were listed but actually only 17 names appear. Obviously the reigns of Zimri (7 days) and of Omri (12 years)[21] were contracted into "Zimri 12 years" (1. 5). Another omission concerns 16 years for the reign of Jehoahaz and Jehoash (1. 10).

The total for the kingdom of Samaria is given in our text as $257^y6^m$ but the given numbers result only in $235^y7^m$. The biblical total would be $241^y7^m7^d$. Gelzer makes the strange statement that the total $303^y7^m$ mentioned in his text is correct; in fact addition results in $242^y$ and some months.

Our text concludes with the remarks:

| | |
|---|---|
| Kings of Samaria ruled until Ḥezekiah | $290^y$ |
| Captivity (of Israel) in Persia | 125 |
| Tribe Judah in Persia | 70. |

These dates agree with the biblical tradition which puts the fall of Samaria in the time of Hezekia of Judah.

Our text contains also a list of "Kings of Edom" which is taken from Gen. 36, 31—39 (cf. Table 8). Neither version contains regnal years or other chronological data.

---

[21] II Kings 16, 15 and 16, 25.

*Table 7*

| | Ethiopic **26** | | Gelzer I p. 99 | | | Bible | | |
|---|---|---|---|---|---|---|---|---|
| 1. | 'Iyorbeʿām, s. o. Nābāṭ | $24^y$ | 22 | 22 | 22 | 22 | Jeroboam, s. o. Nebat | 1. |
| | Nābāṭ, h. s. | 2 | 2 | 2 | 2 | 2 | Nadab | |
| | Baʿas | 24 | 24 | 24 | 24 | 24 | Baasha | |
| | 'Ēla | 2 | 2 | 2 | 2 | 2 | Elah | |
| 5. | Zenberi | 12 | $7^m + 12^y$ | $7^m + 12^y$ | 12 | $7^d + 12^y$ | Zimri + Omri | 5. |
| | 'Aka'ab | 22 | 22 | 22 | 22 | 22 | Ahab | |
| | 'Akāzyās | 2 | 2 | 12 | 2 | 2 | Ahaziah | |
| | 'Iyorām | 12 | 12 | 12 | 12 | 12 | Jehoram | |
| | 'Iyu | 28 | 28 | 28 | 28 | 28 | Jehu | |
| 10. | 'Iyo'akaz | 17 | 17 + 16 | 17 + 16 | 17 + 16 | 17 + 16 | Jehoahaz + Jehoash | 10. |
| | 'Iyorbeʿām | 41 | 41 | 41 | 42 | 41 | Jeroboam | |
| | 'Azāryās | $6^m$ | $6^m$ | $6^m$ | $6^m$ | $6^m$ | Zechariah | |
| | Sēlom | $30^d$ | $1^m$ | $1^m$ | $1^m$ | $1^m$ | Shallum | |
| | Menāḥē | 10 | 10 | 10 | 10 | 10 | Menahem | |
| 15. | Fāqēsyās | 2 | 2 | 2 | 2 | 2 | Pekahiah | 15. |
| | Fāquḥē | 28 | 20 | 20 | 20 | 20 | Pekah | |
| | Hosēʿe | 9 | 9 | 9 | 9 | 9 | Hoshea | |
| | total | $257^y6^m$ | $303^y7^m$ | $303^y7^m$ | | | | |
| | Σ | $235^y7^m$ | $242^y2^m$ | $242^y2^m$ | $242^y7^m$ | $248^y7^m$ | | |

*Table 8*

| **26** | Gen. 36 |
|---|---|
| Bālāq, s. o. Bēʿor | Bela, s. o. Beor |
| 'Iyobāb, s. o. Zārā | Jobab, s. o. Zerah |
| | Husham the Temanite |
| 'Adād, s. o. Bārād | Hadad, s. o. Bedad |
| 'Asmā | Samlah |
| Sa'ol, s. o. Robat | Shaul of Rehoboth |
| Bala'inon | Baalhanan, s. o. Achbor |
| | Hadar |

## *E. Summary*

Looking at the material assembled here, one finds a variety of dates for every major historical event. Nevertheless, some dates show a much higher frequency than the rest and therefore can be considered as representing a widely accepted chronological framework. These dates are:

| | W | Δ | | | W | Δ | |
|---|---|---|---|---|---|---|---|
| Birth of Noah | 1656 | | (8) | David | 4447 | 694 | (8) |
| Flood | 2256 | 600 | (15) | Nebuchadnezzar | 4916 | 469 | (9) |
| Tower | 2827 | 571 | (9) | Birth of Christ | 5500 | 584 | (16) |
| Birth of Abraham | 3328 | 501 | (9) | Diocletian 0 | 5576 | 276 | (16) |
| Exodus | 3753 | 425 | (8) | | | | |

where the numbers in parentheses give the number of tabulations in which these dates are listed. Obviously, the Deluge, the birth of Christ and the era Diocletian are the essential fixed-points of this whole chronology. Yet only a minority of texts adhere strictly to these dates while most lists show more of less random deviations.

This is illustrated in Table 9 which compares four short lists of the dates of major events but disagree in almost all cases. Cf., e.g., the dates 3754, 3816, 3836, 3844 for the Exodus or 5488, 5500, 5562 for the birth of Christ.

*Table 9*

| No. | | 31 Σ | 51 52 Σ | 53 Σ | 54 Σ | | No. |
|---|---|---|---|---|---|---|---|
| 1. | Adam→Flood | 2257 | 2068 | | 2256 | Flood | 1. |
| | Flood→birth of Abraham | 1072 [3329] | | | | birth of Abraham | |
| | Flood→Abraham 75 | | 1256 [3324] | | | Abraham 75 | |
| | Adam→birth of Lewi | | | 3556 | | birth of Lewi | |
| 5. | birth of Lewi→birth of Moses | | | 180 3736 | | birth of Moses | 5. |
| | Flood→Exodus | | | | 1588 [3844] | Exodus | |
| | birth of Abraham→Exodus | 507 3836 | | | | Exodus | |
| | Abraham 75→Exodus | | 430 [3754] | | | Exodus | |
| | birth of Moses→Exodus | | | 80 3816 | | Exodus | |
| 10. | Exodus→Saul | | | | 539 [4383] | Saul | 10. |
| | [Adam]→death of David | | | 4495 | | David | |
| | Exodus→Temple | | 440 [4194] | | | Temple | |
| | Saul→Captivity | | | | 512 [4895] | Captivity | |
| | Temple→Captivity | | 934 [5128] | | | Captivity | |
| 15. | Captivity→Return | | | | 70 [4965] | Return | 15. |
| | death of David→Olymp. 1,1 | | | 250 4745 | | Olympiad 1,1 | |
| | Return→Alexander | | | | 206 [5171] | Alexander | |
| | death of David→death of Alex. | | | 428 5173 | | Alexander | |
| | death of Alex.→death of Cleop. | | | 284 5457 | | Cleopatra | |
| 20. | Alexander→Augustus | | | | 277 [5448] | Augustus | 20. |
| | Captivity→birth of Christ | | [434] 5562 | | | birth of Christ | |
| | Adam→birth of Christ | | 5562 | | | birth of Christ | |
| | death of Cleop.→birth of Chr. | | | [43] 5500 | | birth of Christ | |
| | Augustus→birth of Chr. | | | | 40 5488 | birth of Christ | |
| 25. | birth of Chr.→Crucifix. | | | 33 5533 | | Crucifixion | 25. |
| | Adam→Ascension | | 5594 | | | Ascension | |
| | Resurrect.→D [0] | | | 243 5776 | | D 0 | |
| | Adam→D 1 | | 5860 | | | D 1 | |
| | D [0]→D 76 | | | 76 5852 | | D 76 (= G 0) | |
| 30. | birth of Christ→Islam | | | | 614 6102 | Islam | 30. |
| | Adam→Deḥam 28 | | 6177 | | | Deḥam 28 | |

A unique entry is in **53** the date W 4745 (= 748 B.C.) for the "first Olympiad" (walâpayâdos). In Byzantine sources the Olympiad 1,1 can be associated with at least one of the following dates:

| | | |
|---|---|---|
| King Uzziah 42 = W 4738 | Gelzer, Afric. II, p. 349 |
| Uzziah 51 = 4732 | II, p. 146/7 |
| Ahaz 1 = 4727 | I, p. 97/8 |
| Ahaz 11 = 4745 | II. p. 319 |

According to modern chronology, Ol.1.1. would be W 4716[22], i.e. 776 B.C.

To the lower part of **52** exist two close parallels:

| From Adam to | **52** | **2, 35** | Δ |
|---|---|---|---|
| Birth of Christ | 5562 | 5536 | 26$^{y}$ |
| Ascension | 5594 | 5569 | 25 |
| Diocletian 1 | 5860 | 5852 | 8 |
| Ḍeḥam 28 | 6177 | 6177 | 0 |

The year 5852 is actually the year 0 of the era G, not the year D 1 (a common mistake)[23]. In **53** *C* 5852 is correctly equated with D 76. The last name is known for the 13th king after the Conversion of Ethiopia but raises several chronological questions[24].

## 3. Special Periods

What we are going to discuss in the following comes from a fragment of a chronological work which has no parallel in my material. This text was written on some empty pages (2$^{a}$ and 7$^{b}$, 8$^{a}$ lower half) of the otherwise unrelated large treatise *BN 160.*

The first four sections (from fol. 2$^{a}$) concern: *A* the "Persian" rulers (cf. **20**), *B* Alexander and the myth of the "Bicornute", *C* the Ptolemies (**21**), *D* the Romans (from Augustus to Vespasian, cf. **22**). Between *B* and *A*, as well as between *D* and *C*, a line of separation

---

22 Cf., e. g., Ideler, Chron. I p. 376.

23 Cf. above p. 28.

24 Cf. below p. 65.

is drawn in the manuscript, but not between *B* and *C*. Apparently Alexander and the Ptolemies were considered as belonging to the same historical period.

The Kings of Aksum are treated in a similar fashion in the text preserved on fol. $7^b$ and $8^a$ for which cf. below p. 58 and **61** to **65**. A few short sections from the original text of *BN 160* (cf. **66**) that also concern Ethiopic chronology are independent of **61** to **65**. The same holds of **23** in relation to **20** to **22**.

### *A. The Persians*

Devoting a special section to the Persians (cf. Table 10, p. 45) is, of course, motivated by the role the Achaemenide kings were playing in Jewish history: the return from captivity and the rebuilding of Jerusalem and the Temple. Hence it is not surprising to find, at the beginning of this section, a reference to the 70 years of captivity and the names of the Babylonian kings[25] who ruled during this period. The dates, however, are quite arbitrarily chosen, only such as to reach the required total: $26+23+21=70$. The names

Nabukadanasor
Sêrêlyâl-Mâdâroq
Bêlṭasor (and) Sêyêl-Mâdâroq

are garbled and probably stand for

| | |
|---|---|
| Nebuchadnezzar [Nabu-kudurri-uṣur], | actually 42 or 43 years (605/4 B.C. to 562) |
| Evil-Merodach [Amil-Marduk] | 2 years (561 and 560) |
| Neriglissar[26] [Nergal-šarra-uṣur] | 3 years (559 to 556) |
| Nabonid[27] [Nabu-na'id] | 16 or 17 years (556/5 to 539). |

From the total of 65 or 66 years, 19 years should be subtracted for the time before the fall of Jerusalem in Nebuchadnezzar's reign. Thus there remain only less than 50 years for the Exile.

---

[25] In modern terminology: the Neo-Babylonian dynasty.

[26] Baltasar for Neriglissar occurs frequently in Byzantine chronographies (e.g., Gelzer, Afric. I p. 101, II p. 148, p. 390).

[27] Not again Evil-Merodach, as our text seems to imply.

*Table 10*

*BN 160* 2ª.1—11: *The Persians*

| | | | | Σ | |
|---|---|---|---|---|---|
| after capture Israel remained in Persia 70$^{y}$: | | | | | Captivity: |
| Nābukadanaṣor kept them | | 26$^{y}$ | | | under Nebuchdnezzar |
| Sērēyālmādāroq reigned | | 23 | | | Evil Merodach |
| Belṭāsor (and) Sēyēlmādāroq reigned 21 | | | | | Neriglissar and Nabonid |
| Dāreyos, s. o. 'Aḥesurs of Māhi, and Kuerš of Persia ruled | | | 9$^{y}$ | 9$^{y}$ | Darius I and Cyrus |
| of which both of them (ruled) | | 7$^{y}$ | | | |
| after the death of Dāreyos Qiros | reigned alone | | 3 | 12 | |
| Faḥsayos, s. o. Qiros | reigned | | 8 | 20 | Cambyses?, s. o. Cyrus |
| Dāreyos, s. o. Yesaṣef | reigned | | 3 | 23 | |
| Dāreyos Masaglāy | reigned | | 13 | 36 | Darius I |
| 'Asasurs, h. s. | reigned | | 20 | 56 | Ahasuerus (= Xerxes I) |
| 'Azdārēs | reigned | | 40 | 96 | Artaxerxes I |
| Na?'edāsēr, the second | reigned | | 5 | 101 | Xerxes II |
| Ṣāʿerinos | reigned | | 3 | 104 | Sogdianos |
| Dāreyos, s. o. 'Amat, called Mantu | reigned | | 16 | 120 | Darius II |
| 'Azdeyāsēr, h. s., brother of Qiros | reigned | | 14 | 134 | Artaxerxes II |
| 'Abdāsēr, called 'Akuš | reigned | | 20 | 154 | Artaxerxes III Ochus |
| 'Arsēs, h. s. | reigned | | 4 | 158 | Arses |
| Dārā, s. o. 'Arses | reigned | | 20$^{y}$6$^{m}$ | 178½ | Darius III |
| total | | | 276$^{y}$[1] | | |

[1] Seribal error for 176

The beginning of the list of Persian kings makes no sense as it stands. But the total of 36 years from Darius (s.o. Ahesurs) to Darius (Masaglay) probably represents the length of 36 years commonly assigned to Darius I, preceded by the dynastic struggles which filled much of the reigns of Cyrus and Cambyses.

After Darius I our text gives a correct list of the Achaemenid Kings[28]. There are difficulties, however, with the numbers of regnal years. Sogdianos (Ṣâʿerinos) with 3 years is probably an error for 3 months[29]. For Artaxerxes II our text gives about 30 years too little, an error a scribe tried to repair by giving the last king, Darius III, 20 years in order to restore a reasonably correct total.

There can be no doubt that our Ethiopic chronological sources go back to Byzantine chronographers of different periods. One could hope to find a more accurate relationship of such specific texts as *BN 160* **20** by comparing its data with the data assembled, e.g., in Gelzer's work on Africanus. Table 11 shows the result of such a survey: the discrepancies between the different lists are so great that one cannot hope to establish something like an "archetype" for this kind of historiographic material.

### *B. Alexander the "Bicornute"*

This section (2[a], 11—20) contains only two chronological information. First: Alexander's reign lasted 12 years, and second: "he appointed 4 successors, 1 from Egypt, 1 from Rome, 1 from Persia, 1 from Syria, and he ruled, including (mesla) his successors, 56 years". This makes no sense since, for whatever chronology one assumes, the year Alexander 56 reaches only into the reign of Ptolemaios Philadelphos.

The main part of this section tells about the fight between the "Two-horned" goats that represent Darius III and Alexander, apparently based on Daniel 8. The structure of the text suggests an underlying poem.

---

[28] Cf. CHI 2 p. 874.

[29] For Sogdianus (usually accounted for with 7 months) cf. Weissbach in Pauly Wissowa II, 5 col. 791—793 (1927). Cf. Also Table 11 p. 47.)

*Table 11*

| | Gelzer. Afr. | I p. 104 | II p. 13 | II p. 15 | II p. 49 | II p. 115 | II P. 148/9 | II p. 351 | Ethiopic | modern | B. C. | |
|---|---|---|---|---|---|---|---|---|---|---|---|---|
| 1. | Cyrus | $30^y$ | $30^y$ | $30^y$ | | $31^y$ | | | | | | 1. |
| | Cambyses | 8 | 19 or 9 | 9 | | 9 | | | | | | |
| | Darius I | 36 | 36 or 23 | 36 | $36^y$ | 36 | 36 | 36 | $36^y$ | $36^y$ | 522—486 | |
| | Xerxes I | 20 | 26 or 24 | 26 or 24 | 20 | 21 | 28 | 20 | 20 | 20 | 485—465 | |
| 5. | Artabanes | $7^m$ | | | | | | $7^m$ | | | | 5. |
| | Artaxerxes I | 41 or 40 | 36 or 30 | 36 | 41 | 41 | 41 | 40 | 40 | 40 | 465—425 | |
| | Xerxes II | $2^m$ | $12^y$ or $2^m$ | | | $2^m$ | 14 | $2^m$ | 5 | 1 | 425/4 | |
| | Sogdianos | $7^m$ | $17^y$ or $7^m$ | | | $7^m$ | 7 | | 3 | 1 | 424 | |
| | Darius II | 19 | 18 | 18 | 19 | 19 | | 19 | 16 | 19 | 424—405 | |
| 10. | Artaxerxes II | 42 | 62 or 61 | 62 | 40 | 62 | 40 | 41 | 14 | 46 | 405—359 | 10. |
| | Artaxerxes III | 26 or 22 | $23^y7^m$ or $23^y$ | 23 | 26 | 23 | 27 | 26 | 20 | 21 | 359—338 | |
| | Arses | 4 or 2 | 3 | 3 | 4 | 3 | 4 | | 4 | 2 | 338—336 | |
| | Darius III | $6^y6^m$ or $6^y$ | 12 or 7 | 7 | 6 | 4 | 6 | | $20^y 6^m$ | 5 | 336—331 | |
| | total | $\leqq 196^y$ | $\leqq 246$ | ~210 | 192 | 210 | 203 | 183 | 176 | 191 | | |

### C. *The Ptolemies*

Alexander and Cleopatra[30] are frequently mentioned as the endpoints of the hellenistic period but we have only one text (**21** from *BN 160*) that gives a list of Ptolemaic Kings (cf. Table 12).

The first name is perhaps Claudius[31], referring to the famous astronomer who in the Middle Ages was frequently assumed to have been a King of Egypt. Names of the subsequent rulers composed with "Philos" are translated as mafqarê, "lover" (of brother, father, mother respectively). The years of their reigns listed in the right half of Table 12 are taken from T. C. Skeat's reconstruction of the Ptolemaic chronology. The total of 293 years refers to the whole interval from Philip Arrhidaeus to the death of Cleopatra.

Comparison with Gelzer's material shows reasonably good agreement in the sequence of names but the regnal years differ drastically, particularly at the beginning and near the end of the Ptolemaic period. The following Table 13 shows only the period of relative consistency in Gelzer's data where they are related to our text.

In **3** and **36** the time from Alexander to Cleopatra is given as 294 years in agreement with modern chronology. From **5**, **30**, and **38** in which "Ptolemy 16" (most likely Philadelphus) is equated to 10 cycles, it would follow that his first year would be W 5305 = −187 A.D. (instead of −285 in modern chronology).

### D. *The Romans (from Augustus to Titus and beyond)*

For an Ethiopic author Roman history was of interest only so far as it was directly related to Christianity, hence the time from Augustus[32] to Vespasian and Titus. This period is represented in

---

[30] Her name has suffered a variety of spellings, e.g., 'eklâ'ubaṭra walata deyoryos (i.e. "daughter of Dionysos") or kâleʿe badarâ gebṣâwit ("Cleopatra the Egyptian").

[31] The text adds "also(?) called Gaber (Slave(?)) in Egypt and Alexandria, but Ebaguera by(?) the Arabs".

[32] No definite dates for the reign of Augustus are found in our material. The text **54** would lead to 5448 for "Augustus" but its dates are obviously in disorder. From **22** (cf. Table 15) one would obtain W 5477 for Augustus 1, i.e. B.C. 32, which is essentially correct. In **44** the "beginning of the reign" of Augustus is given as W 5484.

*Table 12*

| baṭlimos | (Ptolemaios) | years | modern | years |
|---|---|---|---|---|
| gleyādiqos | (Claudius?) | 7 | Philip Arrhidaeus | 7 |
| 'eskederos | (Alexander) | 24 | Alexander (II) | 12 |
| 'arsāb? | | 24 | Ptolem. Soter | 20 |
| mafqarē 'eḫuhu | (Philadelphus) | 16 | Philadelphos | 38 |
| ta'azūzi | (Euergetes) | 5 | Euergetes | 25 |
| mafqarē 'abuhu | (Philopator) | 17 | Philopator | 19 |
| baṣuḥ | (Epiphanes) | 24 | Epiphanes | 24 |
| mafqarē 'abuhu | (Philopator) | 25 | Philometor | 35 |
| zagaber | (Euergetes?) | 20 | Euergetes | 29 |
| madḫen | (Soter) | 18 | Soter | 36 |
| mafqarē 'emu fasqos | (Philometor Physkos) | 10 | | |
| qasa'os | (Auletes) | 18 | Auletes | 22 |
| yenāsayos | (Dionysos) | 24 | Dionysos | 29 |
| kale'o baṭarā | (Cleopatra) | 30 | Cleopatra | 21 |
| total | | 262 | total | 293 |

five of our texts: two pairs of almost duplicates (**14**, **45** and **11**, **13** 3) and one slightly longer text **22**, written in Islamic times.

The shortest list (cf. Table 14, from **14** and **45**) dates the destruction of the Temple to the year J 77 and the Ascension to J 31. The longer version (**11** and **13** 3) gives also dates for the writing of the Gospels:

| | |
|---|---|
| Matthew | Claudius 1 |
| Mark | Claudius 4 |
| Luke | Claudius' last year [14] |
| John | Nero 8. |

*Table 13*

| | Ethiop. | I 269 | 273 | 274 | II 17 | | 19 | 142 | 245 | 320 |
|---|---|---|---|---|---|---|---|---|---|---|
| Lagos | | 40 | 42 | 40 | 42 | 52 | | 40 | | |
| Arrhidaeos | 7 | | | 6 | 7 | 7 | 7 | | | 7 |
| Philadelphos | 16 | 38 | 37 | 37 | 38 | 38 | 38 | 38 | 38 | 38 |
| Euergetes | 5 | 24 | 25 | 25 | 30 | 30 | 25 | 26 | 24 | 25 |
| Philopator | 17 | 17 | 17 | 17 | 17 | 17 | 17 | 17 | 17 | 17 |
| Epiphanes | 24 | 24 | | 24 | 23 | 23 | 23 | 24 | 24 | 24 |
| Dionysos | 24 | | 29 | 29 | 29 | 29 | 29 | 30 | 30 | 29 |
| Cleopatra | 30 | | 22 | 22 | 25 | 25 | 24 | 22 | 22 | 22 |

*Table 14*

| | | |
|---|---|---|
| Augustus, lived after the birth of Christ | | $15^{y}$ |
| Tiberius | ruled | 23 |
| Gaius, h. s. | | 4 |
| Claudius, h. s. | | 14 |
| Nero | | 13 |
| three rulers | | 2 |
| Vespasian | | 9 |
| Destruction of the Temple in his 6th year, i. e. the 77th year from the birth of Christ, and the 46th year after the Ascension | | |

The destruction of the Temple is assigned the date Vespasian 6 = $40^{y}$ after Ascension = W 5574.

The fifth list (**22**) adds under Augustus and Tiberius dates concerning Christ which we shall discuss in the next section (p. 52). The name "Augustus" is explained as meaning "splendor" (ḍadâl) and Augustus is named "son of Manuhos" (which I cannot explain) and he is said to have ruled $52^{y}6^{m}$. At the end Galba (gâbyos), Otho (ʾaynun), and Vitellius (faṭelos) are given $9^{m}$, $3^{m}$, $3^{m}$ respectively.

The destruction both of the first and of the second Temple, in W 4905 and 5577 respectively, became epoch dates for Jewish eras. A detailed description of the relevant parameters is given in *BM Add 16252*, Chapter 55, § 4 to 6, in a supplement to Abu Shaker's Chronography (for which cf. Neugebauer [1988] p. 161).

A short section[33] (without parallel) in *BMA 16217* ($22^{a}$ I, 1—13) gives [Decius (?)] 35 years, Diocletian 21, Maximianus 19, Constantine 19. These numbers are then combined with other numbers (30, 70, 100, etc.) leading somehow to a total of 276 years which probably refers to the years from Christ to Diocletian. I do not understand the connections between all these numbers, not even from a purely arithmetical viewpoint.

In three closely related texts, **2**, **35** and **52**, we find the statement that Diocletian ruled with his son Maximianus 28 years, that Constantine's reign lasted 40 years and that the Council of Nicaea was held in his 10th year. No absolute dates for these events are given.

## *E. Christianity, Church Councils*

Our chronographic lists reveal an unexpectedly complex multiplicity for the "Incarnation" (i.e. conception or birth) of Christ beside the customary date W 5500. Table 15 illustrates this situation, showing, e.g., three sets for the birth: beside the "standard" W 5501 IV 29 Tuesday, an earlier date W 5464 IV 28 Tuesday, as well as a later date W 5536, attested in six of our sources:

We are facing here three different chronologies which one could be tempted to describe as using as epoch dates

[33] In part Amharic. Incorrectly described by Dillmann (BM Cat. p. 45) as concerning "historiam creationis".

5464 = 5500 − 36, 5500, and 5536 = 5500 + 36. Unfortunately, this symmetry is not securely established. For the earlier date, e.g., one could also use Baptism and Crucifixion and thus obtain −38 years instead of −36, while the later dates lead to +35. An interval of 38 = 2·19 years has perhaps a parallel in the relationship between the "Spanish era" and the era "A.D."[34]. On the other hand we shall show that 5536 is connected in a more complicated fashion with the specific chronology of Ethiopic history (cf. below).

The importance of the lunar phases for the Easter-dates explains the inclusion of lunar dates (cf. Table 15) both from the Jewish luni-solar calendar and from the Islamic rotating calendar (extended schematically back from the era Hijra). These two calendars coincide excepting for a cyclic deviation of the month-names whereas the day-numbers are kept the same for both calendars[35].

In **22** the year 5500 is equated with the year 34 of the reign of Herod[36]. Dates for the writing of the Gospels (in the traditional order Matthew - Mark - Luke - John) are given in **11** and **13** 3[37]. Even in the standard chronology some smaller discrepancies exist. For example, according to **22**, the Crucifixion would fall into the year Tiberius 18 but according to **10** and **13** 2 in Tiberius 19.

Serious difficulties are found in some short texts (cf. Table 16): the duplicates **2** and **35** start with the year 5536 for the birth of Christ, while **51**, continued in the Ethiopic kinglist **72**[38], assumes 5562 as the year of Christ's birth. But both **2, 35** and **52** jump from Constantine to Ṣaḥam, followed by 5 years for Amida (two Ethiopic kings also listed in **72** and **62** *C*[39]. Even if we accept 5536 and 5562 as two legitimate chronologies for the life of Christ, the common endpoint 6177 for Ṣaḥam 28 does not explain how intermediary events were dated in these two versions. As we shall see in the

---

[34] Cf. for this problem EAC p. 126 and [1981] p. 177/8.

[35] Therefore **45** (cf. Table 15) gives only day numbers without month-names. Day 9 for the day of birth is perhaps a scribal error for 10. Cf. my study on Abu Shaker [1987] p. 64.

[36] Abu Shaker assumes Herod 33 as the year of the birth of Christ (cf. Neugebauer [1988] p. 132). Modern dates: 37 to 4 B.C.

[37] Cf. above p. 50.

[38] Cf. below p. 130.

[39] Cf. also below p. 62.

*Table 15*

| | *BN 160* 24ᵃ/25ᵇ | **10, 13** 2, **44** | | lunar dates | | | | **1, 2, 19** 2 **33—35** |
|---|---|---|---|---|---|---|---|---|
| | | | | **22** | | | | |
| Conception | 5463 VII 29 Sun | 5500 VIII 29 Sun | | Jew. | W′ Islam | 5567 day | 1 | |
| | | | Augustus 25 = | | | | | |
| Birth | 5464 IV 28 Tue | 5501 IV 29 Tue | 550[1] IV 2[9] | X′ 10 | 5[6]69 VIII′ 10 | 5668 | 9 | 5536 |
| Baptism | 5493 V 11 Tue | 5531 V 11 Tue | | | | 5697 | 22 | 5566 |
| | | | Tiberius 18: | | | | | |
| Crucifix. | 5496 VII 27 Fri | 5534 VII 27 | VII 27 | I′ 15 | | 5700 | 15 | |
| Resurrect. | | VII 29 | | | 5[70]3 XI′ 17 | 5700 | 17 | |
| Ascension | | IX 8 | | | | | | 5569 |

*Table 16*

| | A. D. | W | | **1** | **19** 2 | **2, 35** | **51, 52** | **72** |
|---|---|---|---|---|---|---|---|---|
| Christ, Birth | | 5500 | | [5536] | [5536] | 5536 | 5562 | |
| Ascension | | | | | 5569 | 5569 | 5594 | |
| Diocletian 1 | 284 | 5776 | | | | 5852 | 5860 | |
| last year | 305 = Diocl. 22 | | Diocl. 28 | | | [5879] | | |
| Constantine 1 | 306 | | | | | [5880] | | |
| Nicaea | 325 = Const. 20 | 5835 | Const. 10 | | | [5889] | | |
| last year | 337 = Const. 32 | | Const. 40 | | | [5919] | | |
| | | | Ṣaḥam 28<br>Amidā 5 | | | 6177 | 6177 | 6177 |

following, the same question arises also in another area of Ethiopic chronography, only in a much more drastic scale (cf. below p. 56).

But even in a simple and easily accessible case, the dates of the great Church Councils, we find the sources grouped around sharply distinct versions. The following two versions are attested:

| Council | A.D. | Version 1 | Version 2 | Δ |
|---|---|---|---|---|
| Nicaea | 325 | W 5817 | W 5835 | 18 |
| Constantinople | 381 | 5873 | 5893 | 20 |
| Ephesus | 431 | 5923 | 5948 | 25 |
| Chalcedon | 451 | 5944 | 5969 | 25 |

The first version, giving the historically correct dates, is found in the texts **8**, **15**, **40**, and **46**. The second version is given in **6**, **12**, and **13** 4, **39**, for Nicaea and Constantinople in **7**, for Nicaea alone in **19** 1, **31**, **32**. Note that the intervals in Version 2 are not the same as in the correct version 1, a fact for which I have no explanation. In **2**, **35**, **52** the Council of Nicaea is associated with Constantine year 10, in **12** with his 12th year.

The texts **12** and **13** 4 (duplicates) contain many additional information about these Councils, their membership and their topics of discussion[40]. The year 5893 (Council of Constantinople) is also characterized as the 41th year after the completion of the 11th lunar cycle ($11 \cdot 532 + 41 = 5893$) which relates Version 2 to a definite chronological framework[41]. This shows that the dates of Version 2 represent an intentional deviation from the historical data of Version 1, excluding the assumption of simple scribal errors.

In turning now to the chronology of Ethiopic dynasties we shall meet a most unexpected duality: the suppression or acceptance of a whole 532-year cycle.

---

[40] Cf. the translation in Weld Blundell, Roy. Chron. p. 498.

[41] Cf. above Table 1 (p. 29).

## 4. Ethiopic Kings

Many chronological lists that concern Biblical and early Christian history continue with Ethiopic dates, often from the "Conversion" (in the third century) to dates near modern times[42]. The most frequently mentioned fix-points are:

Gabra Masqal (of Aksum)[43]
Zague-dynasty (about 1140—1270)[44]
Yekuno Amlâk (1270—85)
Takla Hâymânot (I: 1706—8).

It is, however, only beginning with Yekuno Amlâk that a fairly accurate modern Ethiopic chronology has been established. The following discussion will show the problems with which we are confronted in consulting the original sources.

Of central importance is the fact that one finds for many Ethiopic reigns two widely differing dates, always 456 years apart[45]. For example Gabra Masqal is given the date W 5929 in the texts **9**, **19** 1, **31**, **32** (and probably also in **33** and **71**) while W 6384 = 12 C is represented in four cases: **1**, **19** 2, **34**, **35**[46].

But this dichotomy in royal chronology reaches deeper, although involving different amounts. The texts with the lower date for Gabra Masqal (5929) assume the commonly used date W 5500 for the birth of Christ, while the later date (6384) is associated with W 5536 for Christ's birth[47].

Even the Church Councils show the influence of this grouping in so far as the "second" (unhistorical) version (cf. p. 54) is used in **19** 1, **31**, **32** and thus with the earlier date (5929) for Gabra Masqal and W 5500 for Christ.

In spite of this clear evidence for two different versions of Ethiopic chronology, beginning with the time of Christ, there is no

---

[42] On the "Solomonic" chronology, invented by Yesḥaq of Aksum in the 14th century, cf. Ullendorff in S. Pritchard, Solomon and Sheba, p. 104.

[43] Conti Rossini, Storia p. 159, suggested a date around A.D. 520.

[44] Tamrat, Church and State, p. 54 ff.; Ullendorff, Eth., p. 65. Cf. also below, p. 64.

[45] Cf. the references given in our chronological list on p. 15 f.

[46] We ignore differences of ±1, explicable by the shift between ordinal and cardinal numbers.

[47] Cf. above p. 51.

arithmetical principle which would explain these greatly variable differences. Only the largest difference, that of 456 years, allows a numerical explanation, since $456 = 532 - 76$ where 532 is the well known luni-solar cycle and 76 the interval from the era Diocletian (D) to the era of "Grace" or "Mercy" (G). Hence a reduction of dates by 456 years can be interpreted as the elimination of one 532-year cycle followed by a shift from the Alexandrian era D to the Ethiopic era G (cf. Fig. 1). The very common mixup between the eras G and D (usually both named "years of the Martyrs"[48]) could also explain that both chronologies could be denotes as years "from Adam".

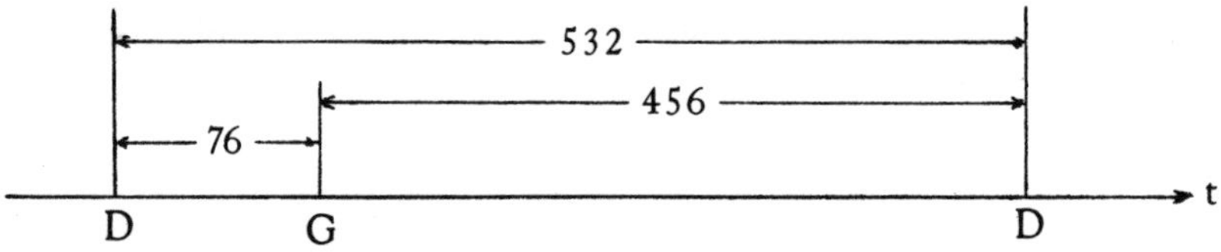


Fig. 1

As a result of this transformation, the dates for the reigns of Gabra Masqal or Yekuno Amlâk and the whole history depending on these dates would appear almost five centuries too early. I am inclined to explain this reduction as a deliberate attempt to hide the fact that the early history of Ethiopia from the time of Aksum and the "Conversion" (around 300) to Gabra Masqal (around 950) had been forgottten (being filled only by the traditional tribal warfare). But the extension of a conveniently shortened chronology far into historical periods (cf. above p. 16: W 7000) is a rather drastic procedure.

In the sources utilized here, one finds no reference to the "era Bizan" which begins 456 years later than the era W[49]. Nevertheless, it can hardly be doubted that our "short" chronology and the "era Bizan" are of common origin. What is new is only the unexpected range in which this era influenced chronographic writing. I see no way to answer the question: What is older, the shortened chronology or the monastery?

---

[48] Cf. p. 28.

[49] Cf. for details EAC p. 123.

## *A. Yekuno Amlâk*

We have two sets of dates for Gabra Masqal, Zague, and Yekuno Amlâk (as well as for many other kings) which differ by 456 years:

| Year 1 of | W′ | W | Δ |
|---|---|---|---|
| Gabra Masqal | 5929 = 11 C + 77 | 6385 = 12 C + 1 | |
| Zague | 6173 + 321 | 6629 + 245 | 244 |
| Yekuno Amlâk | 6306 + 454 | 6762 + 378 | 133 |

The intervals Δ are the same for both versions.

For the year 6762 of Yekuno Amlâk we have (in **56**, **58**, **68**, **70**) the relations

Adam → Yekuno Amlâk 6762
Islam[50] → Yekuno Amlâk 648.

This gives for the first year of the Hijra: W 6114 = J 614 = A.D. 622, i. e., the historically correct date. Hence only the second (later) set must be taken as historically serious[51].

The above named reigns are also related to the Councils of Nicaea (A.D. 325) and Constantinople (A.D. 381) by the statements:

Nicaea → Gabra Masqal: $94^y$ (texts **60** and **67**)
Constantinople → Gabra Masqal: $36^y$ (text **55**).

For these Councils we had found (above, p. 54) the following two versions of dates:

Nicaea: W 5817 or 5835
Constantinople: 5873 or 5893

of which the first one represents the historically correct chronology. The second version, however, leads for Gabra Masqal to the date

$$5835 + 94 = 5893 + 36 = 5929$$

which is the reduced date (W′) for his reign[52]. It is important to see the ambiguity of the Ethiopic royal chronology extends back into the time of the Church Councils.

---

[50] Tanbalât (Dillmann, Lex. 562).

[51] Hence, e.g., Chaine, Chron. p. 246: A.D. 1270 for Yekuno Amlâk.

[52] We have W 6385 in **1**, **19** 2, **34** but W′ 5929 in **19** 1, **31**, **32** and **71**.

### *B. The Kingdom of Aksum; BN 160*

*We have two groups of sources which concern, at least at the beginning, the chronology of Aksum: one from manuscripts in the Bodleian Library, published by Dillmann in 1853*[53]*, the other, quite similar, inserted by a second hand, into BN 160* (fol. $7^{b}$ and $8^{a}$). Conti Rossini included his text in this survey of Aksumite chronology [1909], but neither he nor Dillmann made a serious effort to bring some order into the arithmetical data provided in these texts.

In the following, I am concentrating on the analysis of *BN 160*, a text which I divided into five sections, *A* to *G*[54], which correspond roughly to Dillmann's grouping of his sources. The sections *E* to *G* are closely related to the list of kings that was inserted into the Kebra Nagast.

As usual, our texts show all kinds of minor discrepancies. Nevertheless, there exist also some generally accepted data. In the present context we find three such basic elements:

(1): a mythological period of 800 years for 5 kings, from Arwê (i.e., King "Snake") to Mikida, the Queen of Sheba.

(2): the birth of Christ in the year 8 of king Bâzên.

(3): the interval of 245 years from the birth of Christ to the "Conversion" of Ethiopia in the reign of Abreha and Aṣbeḥa[55], the "loving brothers".

As an example of internal inconsistencies, I quote four intervals which begin with Arwê:

| | | | |
|---|---|---|---|
| **61** *A*: (1) Arwê → Queen of Sheba | | 5 kings | $800^{y}$ |
| (2) Arwê → Bâzên 8 (birth of Christ) | | | $1809^{y}$ |
| | **59** (cf. p. 116) | 25 kings | |
| | Dillmann p. 341 | 21 kings | |
| | Conti Rossini p. 289 | 27 kings | |

---

[53] ZDMG 7 (1853), henceforth quotes as "Dillmann", Bodleian Eth. MSS. 26, 28, 29, 32.

[54] In the manuscript dividing lines are drawn after the sections *C*, *D*, and *F*.

[55] For the significance of the names (henceforth quoted as "A. and A."), cf. Ullendorff, Ethiopia p. 121.

**63** *D*: (3) Arwê → Nalkê (father of Bâzên)[56] 51 kings $859^{y}$
(4) Arwê → A. and A.[57] 85 kings $2104^{y}$

These data do not agree with the totals obtainable from the lists found in the texts, e.g. 44 kings, $1224^{y}$ for (3) and 58 kings $1458^{y}$ for (4). But what is worse: (1) and (3) imply that only 59 years are left between the Queen of Sheba and the time of Christ. This discrepancy, however, can easily be eliminated y emending 859 to 1859 which results also in the proper distance of $245^{y}$ between Nalkê (the time of Christ) and A. and A.

Similarly the interval (2) of $1809^{y}$ can be derived from an interval of $1800^{y}$ to Nalkê, hence of 1000 years from the Queen of Sheba. Thus we obtain the following framework for the chronology of the Kingdom of Aksum:

| | |
|---|---|
| Arwê → Queen of Sheba | $800^{y}$ |
| Queen of Sheba → Bâzên | 1000 |
| Arwê → Nalkê (Bâzên) | 1850 |
| Arwê → Bâzên 8 (Christ) | 1859 |
| Bâzên → A. and A. | 245. |

This reconstruction agrees also with the commonly accepted biblical chronology which assumes for David, year 1, the date W 4447[58]. Hence we may assume for Solomon a date around 4500, i.e. for the Queen of Sheba's visit to Jerusalem about 1000 years before Christ. Hence the only mythological element in this chronology of Aksum is the interval of 800 years between King Snake and Solomon. But all the rest simply reflects the Solomonic tradition of Ehtiopic kingships and may well contain a sound historical core.

BN 160, *A*: $7^{b}$, 1—20 (**61**)

This section, headed "years of the kings of Aksum" is divided into three independent groups: first the mythological interval of 800 years from the King "Snake" to Mikida, the "Queen of the

---

[56] Dillmann p. 342 (B 25).

[57] He is also said to have built Aksum (Dillmann p. 345, 2 B).

[58] Cf. above p. 39. In **53** W 4495 is given as the year of his death, i.e. a little more than 1000 years before the birth of Christ. Cf. also p. 31 for the round date 3300 for Abraham's birth.

South"[59]; then a short list (5 names) beginning with Bayna Leḥkam[60], i.e. Menelik, the son of Solomon and the Queen of Sheba[61]; finally a list of 32 kings and their regnal years, from Baḥas to Nâlkê, the father of Bâzên.

For the first mentioned interval of 800 years we have three close parallels: *BN 160 A, A*̧´ and Dillmann *a* to *f* (p. 341)[62]. In *A* Mikida's visit to Solomon is placed into his 4th year, or Saul's 36th (what does this mean?), while the Kebra Nagast counts her visit as the 6th year of her reign[63]. We are also told that she ruled 25 years after her return from Jerusalem[64].

The next group again concerns 5 rulers, beginning with Leḥkam's reign of 29 years. His successors occupy $4+11+3+44=62$ years; hence we have a total of 91 years for 5 kings. Then follows a short (corrupt?) passage saying that 64 years are the time from Q̂âsyo to Qatr of Mawaṭ[65] without telling us how many kings were involved. But since we have $5+32=37$ names listed for the 45 kings from Leḥkam to Bâzên we can assume 8 kings for the 64 years from Q̂âsyo to Q̣atr.

The next group names 11 kings (in Dillmann 13) for 132 years (117 in Dillmann). Additional 21 kings (20) and 170 years (240?) bring us to the time of Christ. I see no justification for simply disregarding these later lists which reach into the Ptolemaic and Roman period.

At the end of section *A* we return again to the early times but with slightly different intervals:

| | | | |
|---|---|---|---|
| Arwê | → Bâzên 8 | | $1809^y$ |
| Leḥkam | → Bâzên 8 | 45 kings | $984^y$ |

from which would follow Arwê to Leḥkam $825^y$. Since Arwê to Mikida covers 800 years, there would remain 25 years as interval between Mikida and her son. This probably corresponds to the 25

[59] Mâkedâ, Queen of Sheba, passim in the Kebra Nagast.

[60] Dillmann: Ebn-leḥakim = Ibn Hakim.

[61] Kebra Nagast Ch. 32.

[62] This interval is given here as 801 years, a variant without significance.

[63] One MS of the Kebra Nagast gives $50^y$ which makes no sense; but cf. Dillmann p. 341 A note 1.

[64] In *A* and in Dillmann p. 341 A note 1.

[65] Cf. for this passage the note to **61** l. 17 (p. 143).

years which at the beginning of section *A* were assigned to the reign of Mikida.

BN 160, *B*: $7^b$, 20—25 (**62**)

This short section ends with the reign of Aṣbeḥa and Abreha and the statement that the Conversion of Ethiopia took place in the 4th year of their reign. The list of 12 names and regnal years adding up to 193 years should, according to a final statement, cover 13 kings and the canonical 245 years from the birth of Christ. This would account for 193 + 33 = 226 years reckoned from Bâzên 8. Note that section *E* assigns 18 kings to these 245 years.

Comparison with Dillmann's texts only increases the confusion: p. 345 *B* gives for Bâzên to A. and A. 10 (or 11) kings; p. 346 *C* 14 (or 15) kings (no years listed). Completely out of order is p. 343/4 *A* which claims for the same interval 31 (or 32) kings and 425 years.

BN 160, *C*: $7^b$, 25—30 (**62**)

A list of 26 kings, beginning with Abreha and ending with Constantinus (perhaps Constantinus I = Zar'a Yâ'eqob). The regnal years add up to about 240 years. I do not know what dynasty is represented by these kings.

For the first 15 names, ending in Ṣaḥam year 18, we have a parallel list of 13 names in *V 1* **72**, also ending in Ḍeḥam but year 28 (cf. Table 17 p. 62). Other reigns differ also in these two texts, such that Ṣaḥam 18 in **62** would be Abreha 122, but in **72** Ḍeḥam 28 = Abreha 139. Cf. for details below p. 65.

BN 160, *D* and *E*: $7^b$, 30—36 and $7^b$, 36—$8^a$, 21 (**63**)

Here we return to a summary concerning the "kings of Aksum" and the data displayed also in *A* and *B*:

| | | | |
|---|---|---|---|
| Arwê to Nâlkê: | 15 kings | 859 years | |
| Bâzên ruled: | | 18 years | Birth of Christ: year 8 |
| Bâzên to A. and A.: | 18 kings | 245 years | (Conversion) |
| Arwê to A. and A.: | 85 kings | 1104 years | (scribal error: 2104) |

Construction of the Temple ??

Year 426 of Cycle 11 (= 5320 + 426 = 5746, i.e. J 245 = Conversion).

*Table 17*

| | **62** C | | Σ | **72** | | Σ | Dillman [1853] | | Variants | | Σ | Variants | | re-stor-ed | Σ | |
|---|---|---|---|---|---|---|---|---|---|---|---|---|---|---|---|---|
| 1. | Abreha | 10 | | Abrehā | 12 | | Aṣbeḥa ’ela Abreḥa | 12 | | | | | | 12 | | 1. |
| | Asefḫa | 3 | 13 | Afśeha | 7 | 19 | Asfeḥa | 7 | | 5 | 19 | | 17 | 7 | 19 | |
| | Śāhl | 14 | 27 | Śāhl | 14 | 33 | Śāhl | 14 | | | 33 | | 31 | 14 | 33 | |
| | | | | Adḫano | 14 | 47 | Adḫanā | 14 | | | 47 | | 45 | 14 | 47 | |
| 5. | Reteʿe | 1 | 28 | Retaʿe | [1] | 48 | Reteʿe | 1 | | | 48 | | 46 | 1 | 48 | 5. |
| | Êsfeḥ | 5 | 33 | Asfeḥa | 5 | 53 | Asfeḥ | 1 | 5 | | 49 | 54 | 47 | 5 | 53 | |
| | Aṣbeḥa | 16 | 49 | Aṣboḥ | $16^{?}$ | 69 | Asbeḥa | 5 | 16 | 16 | 54 | 70 | 63 | 16 | 69 | |
| | Aminādā | 7 | 56 | Amidā | 7 | 76 | Amēdā | 16 | 6 | 6 | 70 | 76 | 69 | 7 | 76 | |
| | Abrāḫ | $2^{m}$ | | | | | Abrehā | $6^{m}$ | $2^{m}$ | $2^{m}$ | | | | $2^{m}$ | | |
| 10. | Śāhl | $2^{m}$ | | | | | Śāhl | $2^{m}$ | | | | | | $2^{m}$ | | 10. |
| | Gabaz | 14 | 70 | Gabaz | 14 | 90 | Gabaz | 2 | 14 | | 73 | 90 | 72 | 14 | 90 | |
| | Sehul | 4 | 74 | Śehuʿul | $1^{?}$ | 91 | Seḥul | 1 | | | 74 | 91 | 73 | 4 | 94 | |
| | Aṣbeḥa | 3 | 77 | Aṣbahu | 3 | 94 | Aṣbaḥ | 3 | | 2 | 77 | 94 | 75 | 3 | 97 | |
| | Abreḫa za’ela Eder | 17 | 94 | Abrehā | 17 | 111 | Abreh wa’ela Adḫanā | 16 | | | 93 | 110 | 91 | 16 | 113 | |
| 15. | Ṣaḥam | 18 | 112 | Ḍeḥam | 28 | 139 | Ṣaḥam | 28 | | | 121 | 138 | 127 | 28 | 131 | 15. |
| | Amidāb | 12 | | | | | Amidā | 12 | | | 133 | 150 | 139 | 12 | 143 | |

From the above given intervals one should obtain

Arwê to Conversion: 51 + 18 = 69 kings, 859 + 245 = 1104 years

instead of 85 kings in the text.

All this disagrees with the summaries in *A* and *B*. Only the number 18 for the kings from Christ to the Conversion is confirmed in a list of 18 names (and 245 years) in *E* that agrees with the Kebra Nagast.

BN 160, *F* and *G*: 8[a], 21—28 and 29—36 (**64** and **65**)

These two sections concern the Ethiopic Kings from the Conversion to Delna'ad and from Yekuno Amlâk to Yesḥaq. The subdivision given in the text are as follows:

| | | |
|---|---|---|
| Conversion to Gabra Masqal | 9 kings | 184 years |
| Gabra Masqal to Delna'ad | 24 | 244 |
| Zague to Yekuno Amlâk | | 133 |
| Yekuno Amlâk to Sayfa Arʿâda | 8 | 74 |
| Sayfa Arʿâda to Yesḥaq | 4 | 70 |
| Adam to Yesḥaq | | 6400. |

No individual regnal years are given for any of these dynasties which makes it impossible to check the totals.

On the basis of the numbers given here, one would have

from Conversion to Yesḥaq: 705 years

and hence for the Conversion the date W 6400 − 705 = 5695 = J 195, i.e. 50 years less than the usually assumed date J 245. Computing back from W 6400 one would find for Yekuno Amlâk the year 6256, which is again 50 years earlier than 6306, the earlier date found before (p. 57). These data suggest the emendation

Adam to Yesḥaq: 6450[y]

instead of 6400 given in *G*.

This emendation is explicitly confirmed by two texts (**55** and **67**) which give W 6450 as the date for Yesḥaq. Combining this with the interval of 705 from the Conversion gives correctly W 6450 − 705 = 5745 = J 245 as date of the Conversion.

The above given list tells us furthermore that 74 + 70 = 144 years separate Yekuno Amlâk from Yesḥaq. Hence we obtain for

Yekuno Amlâk the date 6450 – 144 = 6306 which is indeed the date for Yekuno Amlâk according to the shorter chronology (cf. p. 57). The same result is obtained for Gabra Masqal and for Zague because the intervals shown in our present list, 244 and 133 years, are the same as found before (p. 57).

## The Zague Dynasty

According to Ethiopic tradition, the Solomonic dynasty was interrupted at the time of king Na'od by a "foreign" rule, the Zague dynasty, and restored to its legitimate branch only much later by Yekuno Amlāk. Modern discussion[66] centers, of course, on the question: how much later?

Unfortunately one has only one chronological list for the Zague dynasty, published by Dillmann in 1853 (p. 350 f.). It gives the names and regnal years of 11 kings, of which nine in a row are assigned 40 years each[67]. Intervals of 40 years (or $80^{y}$) are not rare in our material[68] for covering inaccurately known intervals, perhaps representing estimates based on "generations". Hence it is clear that this account of the Zague dynasty represents only an estimate of about $350^{y}$ for its duration's[69].

This estimate is supported by our text **69** which assigns $336^{y}$ to the Zague dynasty. But the majority of texts reports much lower values: $250^{y}$ "no kingship of Israel" in **66** 1 and **66** 2, and several times only $133^{y}$[70], a value favored by Conti Rossini and generally accepted.

In the meantime, however, the problem has been further complicated by the discovery of the duplication of Ethiopic chronology[71], involving an amount of $456^{y}$. All cases where $133^{y}$ are mentioned for the Zague dynasty are associated with the shorter chronology, i. e. with an account which eliminated $456^{y}$ from actual history. Obviously, this cutting down of intervals must affect the

---

[66] Cf. for references Tamrat, Church and State, p. 54 ff., p. 66 ff.; also Perruchon, Lalibala, p. II f.

[67] 48 years are given to the 8th king, obviously a contamination of $40^{y}$ and the sequence number 8.

[68] Cf., e. g., **18, 24, 25.**

[69] The totals in Dillmann's MSS vary between 330 and 376.

[70] In **55, 60, 65, 67, 71.**

[71] Cf. above p. 55.

time before Yekuno Amlāk. Hence one must face the possibility that the shortest interval given for the Zague may be influenced by the desire to reach better agreement with the shorter chronology. This would mean that Dillmann's list may reflect the fact of a much longer duration of the rule of the Zague than suggested by the $133^{y}$ dates.

### From "Conversion" to Ṣaḥam 28

As mentioned before, the Conversion of Ethiopia is traditionally associated with the reign of Aṣbeḥa and Abreha, 245 years after the birth of Christ[72] in the 8th year of the reign of Bâzên of Aksum. According to some texts[73] A. and A. ruled together $27^{y}$ $6^{m}$ and, after Abreha, Aṣbeḥa alone 12 years. But **63** *D* and *E* and Dillmann [1853] p. 345 (2A) place the Conversion into year 13 of A. and A., 18 kings after the birth of Christ (in Bâzên 8). In **62** 1. 12, however, A. and A. year 4 is assumed as the year of the Conversion, 13 kings after Bâzên. The version given in **63** *D* seems to me preferable since it preserves some significance for the 12 years mentioned above and since it also agrees with the number of kings listed in **63** *E*.

A fixpoint after A. and A. seems to be the year 28 of a king named Ṣaḥam (or Ḍeḥam), given as W 6177 in the texts **2**, **35**, **52**[74]. In **72** he is the 13th king after the Conversion. Table 17 (p. 62) shows the variations concerning this interval. Nevertheless its length, if derived from the given regnal years, can hardly be much different from 155 years. This would lead for Ṣaḥam 28 to a date near W 5745 + 155 = 5900 instead of 6177.

Roughly the same discrepancy can be deduced from **62** (above p. 62) which equates Ṣaḥam 18 with Abreha 122 while **72** gives for Ḍeḥam 28 the year Abreha 139. Hence for Ḍeḥam 28:

from **62**: J 245 + 132 = J 377 = W 5877
from **72**: J 245 + 139 = J 384 = W 5884

instead of W 6177. I cannot explain this large discrepancy.

In **2**, **35**, **52** Ṣaḥam is followed by one more name "ela Amidâ 5

---

[72] Hence the date of conversion W 5745; cf. above p. 20.
[73] Conti Rossini [1909] p. 292, Nos. 59—60.
[74] Cf. also above p. 43.

(years)". Dillmann [1853] p. 347 Nos. 15 and 16 gives for this ela Amida 12 years, and **62** *(BN 160 C)* assigns to Ṣaḥam 18 years only, followed by Amidâb (sic) with 12 years.

### *C. Royal Names*

The lists **61** and **62** (i. e. *A* to *C* from *BN 160*) associate, with very few exceptions, numbers of regnal years with names that consist only of a single word. In the sections *E* and *F*, however, (i. e. in **63** and **64**) which do not list regnal years, one could be tempted to assume names of two components (like Gabra Masqal or Bâḥra Asgad) were it not that the explicitly mentioned number of kings (e. g. 18 kings in *E* which shows only 22 names) requires single-word names in the majority of cases. For the period before the Zague Dynasty, single words are therefore the rule.

For the time after this "foreign" reign one can notice a drastic change: beginning with Yekuno Amlâk we have consistently two-component names, e. g., 20 names for 10 kings[75].

This simple rule does not agree with the division of names given in the list of kings that intruded at the end of the Kebra Nagast[76]. But this list (which we also know from *BN 160 E* to *G*) gives no years, so that the interpretation of the names depends mainly on the punctuation shown in the manuscripts, a notoriously insecure source of information. Accepting the punctuation would result in reigns of almost 20 years for all rulers, a very implausible average[77]. Hence one should disregard the separation of names as given in the publication of the Kebra Nagast.

Several lists, covering about ten reigns before or after A. and A., prefix almost every name with the word 'ela, or za'ela and la'ela, or za alone[78]. Dillmann takes this 'ela as part of the name, transcribing it as Ela-NN. He is followed by Conti Rossini. I have no explanation for this peculiar terminology and its variants.

---

[75] E. g., in **65** *G*.

[76] Bezold, text p. 173, translation p. 138.

[77] Chaine, Chron. p. 247, names for the 508 years from Yekuno Amlâk to Takla Hâymânot 40 rulers, i. e. an average of about 13 years for each king.

[78] Cf. **62** *B* and *C*, **72**, and Dillmann p. 344 (2A) and p. 346 (3A). Note that 'ela is always separated from the subsequent name.

### *D. Kinglists from Gabra Masqal or Yekuno Amlâk toward modern times*

Some ten of our manuscripts contain lists of Ethiopic kings (usually names, throne names, and years of reign) which are historically attested, beginning with Yekuno Amlâk (end of the 13th century) down to the beginning of the 18th. These lists essentially agree with the dates accepted in modern chronology and need not be presented here.

This does not apply to a group of dates from Gabra Masqal to Yekuno Amlâk that shorten the dates by 456 years as described above p. 56 and is shown by the reference on p. 15 and 16.

*Table 18*

| No. | From | Ending with | Kings |
|---|---|---|---|
| **67** | (Nicaea→) Gabra Masqal (5929) | Zareʿa Yâʿeqob (6471) | 6 |
| **55** | (Constantinople→Gabra M. (5929) | Lebna Dengel (6576) | 25 |
| **60** | (Nicaea→Gabra Masqal (5929) | Fasiladas (6781) | 19 |
| **66** | Gabra Masqal | Del Na'od 663ʸ | 20 |
| **65** | (Zague→) Yekuno Amlâk 133ʸ | Yeshaq (6906) | 15 |
| **68** | Yekuno Amlâk | Yaʿeqob (7099) | 21 |
| **69** | (Zague→) Yekuno Amlâk 336ʸ | Iyo'as (7257) | 38 |
| **56** | (Islam→) Yekuno Amlâk (6762) | Yaʿeqob and Zadengel (7099 + 10) | 20 |
| **57** 1 | (Islam→) Yekuno Amlâk (6762) | Iyasu (7174 + 24) | 32 |
| **57** 2 | (Islam→) Yekuno Amlâk (6762) | Takla Hâymânot (7201) | 30 |
| **58** | (Islam→) Yekuno Amlâk (6762) | Iyoʿâs (7257) | 29 |
| **70** | (Islam→) Yekuno Amlâk (6762) | Yoḥanes (7390) | 38 |
| **71** | Christ→Abreha (6035) | Ḍeḥam 28 (6177) | 14 |
| **66** | Bâzên 8 (= Christ) | Lebna Dengel (6679/6798) | |

The year 7000 of the "Second Coming" is transgressed in several of these lists, Na'od being the king into whose reign this (uneventful) year fell. Nevertheless references to W 7000 are common in computus treatises as well as in the present material (Nos. **55**, **57** 1, **57** 2, **60**), written long after 7000.

Table 18 enumerates the lists that concern historical periods. In **60** a note is added "Abu Shaker's computus 7138", which is a correct date. From **65** it follows Zague 6762 – 133 = 6629, which is correct (Yekuno Amlâk 6762), but from **69** Zague 6762 – 336 = 6426[79].

[79] E. g. in Chaine, Chron. p. 246 f. or Hakluyt Soc., Ser. II No. 107 p. 218 f.

# III. THE TEXTS

**1 to 54: World- and Biblical History**

**55 to 72: Ethiopic Kings**

**Notes to the texts: p. 131**

# World- and Biblical History

## 1: *Berol 84* $7^{b}$ II,24—$8^{d}$ I, 21

| | years | Σ |
|---|---|---|
| Adam→Flood | 2128 | |
| Flood→Tower | 540 | 2668 |
| Tower→Abraham $75^{y}$ | 1071 | 3739 |
| Abraham $76^{y}$→Exodus | 430 | 4169 |
| Exodus→Temple | 440 | 4609 |
| Temple→Captivity | 321 | 4930 |
| Captivity→Ezra | 70 | 5000 |
| Ezra→birth of Christ | 536 | 5536 |
| birth of Christ→year 1 of the Martyrs | 316 | 5852 |
| year 1 of the Martyrs→spread of Christianity | 111 | 5963 |
| spread of Christianity→year 1 of Gabra Masqal | 421 | 6384 |
| year 1 of Gabra Masqal→completion of 13 cycles | 532 | 6916 |
| additional $84^{y}$ to 13 C: total | 7000 | |

## 2: *Berol 84* $8^{a}$ I,22—II,7

| | | years |
|---|---|---|
| Adam→birth of Christ | | 5536 |
| Adam→ascension | | 5569 |
| Adam→Diocletian 1 | | 5852 |
| Diocletian with Maximianus, h. s., | ruled $28^{y}$ | |
| Constantine | ruled 40 | |
| Constantine 10: Council of Nicaea | | |
| Adam→Ṣaḥam, regnal year 28 | | 6177 |
| 'ela Amidā | 5 | |

## **3:** *Berol 84* $8^{a}$ II, 7—24

| "Number of years according to the Computus of the Copts, from our father Adam to the coming of our Lord Jesus Christ, praise to him" | | |
|---|---|---|
| | years | Σ |
| Adam→Noah | 1642 | |
| Noah→Flood | 600 | 2242 |
| Flood→construction of the Tower | 350 | 2592 |
| constr. of the Tower→dispersion of languages | [208] | 2800 |
| [from then]→Lord speaking to Abraham | 240 | 3040 |
| our father Abraham→Exodus | 438 | 3880 |
| from then→Alexander | 1093 | 4973 |
| Alexander→Cleopatra | 294 | 5267 |
| from her→annunciation by Gabriel | 233 | 5500 |
| total | 5500 | |
| incarnation→years of the Martyrs | 276 | 5776 |

## **4:** *Berol 84* $8^{a}$ II, 24—$8^{b}$ I, 12

| "And from another book" | years | Σ |
|---|---|---|
| Adam→Flood | 2256 | |
| Flood→constr. of Tower | 578 | 2834 |
| constr. of Tower→Lord speaking to Abraham | 569 | 3403 |
| our father Abraham→Exodus | 432 | 3835 |
| Exodus→reign of David | 606 | 4441 |
| David→destruction of Jerusalem | 488 | 4929 |
| Jerusalem→reign of Alexander | 263 | 5192 |
| Alexander→end of reign of Cleopatra | 279 | 5471 |
| from then→birth of Christ | 30 | 5501 |
| Christ→Diocletian | 276 | 5777 |
| total | 5777 | |

## 5: *Berol 84* 18$^{b}$ I, 1—II, 3

| "Computus for 13 cycles" | |
|---|---|
| Cycle 1 | Adam→'Ēnos 97 |
| Cycle 2 | 'Ēnos 98→[Yārēd] 104 |
| Cycle 3 | Yārēd 105→Lāmēh 122 |
| Cycle 4 | Lāmēh 123→Noḫ 472 |
| Cycle 5 | Noḫ 473→'Ebēr 128 |
| Cycle 6 | 'Ebēr→Nākor 4 |
| Cycle 7 | Nākor 5→'Embarm 42 |
| Cycle 8 | 'Embarm 43→Gēdeyon, the judge, 1 |
| Cycle 9 | Gēdewon 2→Ḥezeqyās, king of Judah, 22$^{y}$6$^{m}$ |
| Cycle 10 | Ḥezegyās 23½→king Ptolemy 15 |
| Cycle 11 | king Ptolemy 16→Martyrs 76 |
| Cycle 12 | Martyrs 77→Martyrs 608 |
| Cycle 13 | Martyrs 609→Martyrs 1140 |
| | total: 6916 1 2 3 4 5 6 7 |

## 6: *Berol 84* 20$^{b}$, 19—21$^{a}$ I, 28

| | years | W | D | | | |
|---|---|---|---|---|---|---|
| Adam→Flood | 2256 | | | | | |
| Flood→Tower | 544 | 2800 | | | | |
| Tower→Lord appears to Abraham | 640 | 3440 | | | | |
| Lord appears to Abraham →Moses the prophet | 440 | 3880 | | | | |
| Moses→Christ | 1620 | 5500 | | W | J | D |
| Christ→Martyrs | 276 | | | 5776 | 276 | 0 |
| Martyrs→Nicaea | 59 | | | 5835 | 335 | 59 |
| Nicaea→Constantinople | 58 | | 117 | 5893 | 393 | 117 |
| Constantinople→Ephesus | 55 | | 172 | 5948 | 448 | 172 |
| Ephesus→Chalcedon | 21 | | 193 | 5969 | 469 | 193 |
| Chalcedon→Islam | 140 | | 333 | 6109 | 609 | 333 |
| Beginning of World→Islam | 6109 | | | | | |
| additional years | 891 | | | | | |
| completion | 7000 | | | | | |

### 7: *Berol 84* $22^{b}$ I, 2—II, 16

| | years | total | Σ |
|---|---|---|---|
| Adam→Flood | 2256 | | 2256 |
| Flood→Tower | 571 | | 2827 |
| Tower→Abraham | 501 | 3328 | 3328 |
| Abraham→Moses | 4[25] | | 3753 |
| Moses→David, king of Israel | 694 | 4447 | 4447 |
| David→Nabukadanaṣor | 469 | | 4916 |
| Nabukadanaṣor→Alexander | 265 | | 5181 |
| Alexander→birth of Christ | 319 | 5500 = 10 C+180 | 5500 |
| (Christ)→Conversion of Ethiopia | 245 | | 5745 |
| Conversion→Diocletian | 31 | | 5776 |
| Diocletian→Nicaea | 59 | [5835] = 11 C − 17 | 5835 |
| Nicaea→Constantinople | 58 | 5893 | 5893 |
| Constantinople→Gabra Masqal | 36 | [5929] | 5929 |

continued for Ethiopic rulers to Lebna Dengel (W 6576): cf. **55**

### 8: *BM 754* $4^{a}$ I, 1—9

| | years | J | W |
|---|---|---|---|
| Birth of Christ→Conversion | 245 | 245 | 5745 |
| Conversion→Diocletian | 31 | 276 | 5776 |
| Diocletian→Council of Nicaea | 41 | 317 | 5817 |
| Nicaea→Constantinople | 56 | 373 | 5873 |
| Constantinople→Ephesus | 50 | 423 | 5923 |
| Ephesus→Chalcedon | 21 | 444 | 5944 |
| Chalcedon→Islam | 170 | 614 | 6114 |

continued for Ethiopic rulers to Yāʿeqob and Zadengel (J 1599); cf. **56**

## **9**: *BM 815* $17^{a}$ I, 15—$17^{b}$ I, 3

| creation of Adam→Incarnation: 5500 | years | total |
|---|---|---|
| Adam→birth of Noah | 1656 | |
| Noah→recession of Flood | 600 | 2256 = 4C + 128 |
| recession of Flood→Tower | 571 | 2827 = 5C + 167 |
| Tower→Abraham | 501 | 3328 = 6C + 136 |
| Abraham→Moses | 425 | 3753 = 7C + 29 |
| birth of Moses→David | 694 | 4447 = 8C + 191 |
| David→Nābukadanaṣor | 469 | 4916 = 9C + 128 |
| Nabukadanaṣor→'Eskender | 265 | 5181 |
| 'Eskender Maqēdonāwi→birth of Christ | 319 | 5500 = 10C + 180 |

## **10**: *BM 815* $17^{b}$ I, 3—19

| Christ | | |
|---|---|---|
| conception | Magābit (VII) 29, Sunday | |
| birth | Tāḫśāś (IV) 29, Tuesday, reign of Augustus | |
| baptism | Ṭer (V) 11, Tuesday | 5531, Tiberius year 16 |
| crucifixion | Magābit (VII) 27 | 5534 |
| resurrection | Magābit (VII) 29 | |
| ascension | Genbot (IX) 8 | |

| Additional data | | EAC p. 59—61 | |
|---|---|---|---|
| birth | t = 1, e = 9 | k = 181, | c = 10 |
| baptism | t = 3, e = 11 | 211 | 2 |
| crucifixion | e = 14, m = 16 | 214 | 5 |

## 11: *BM 815* $17^b$ I, 20—II, 17

| | | | |
|---|---|---|---|
| Augustus lived after birth of Christ | | | 14 years |
| Tiberius | ruled | | 23 |
| Gaius | | | 4 |
| Claudius | | | 14 |
| Matthew wrote Gospel in year | Claudius | 1 | = after ascension 8 years |
| Mark | Claudius | 4 | 11 |
| Luke | | Claudius last year | 22 |
| Nero | ruled | | 13 |
| John wrote Gospel in year | Nero | 8 | |
| execution of Peter and Paul | Nero | 13 | "on a Sunday" |
| Vespasian | ruled | | 9 |
| Destruction of Temple | Vespasian | 6 | = after ascension $40^y$ = W 5574 |

## 12: *BM 815* $17^b$ II, 17—$18^b$ I, 11

| | years | J | total |
|---|---|---|---|
| Birth of Christ→Conversion of Ethiopia | 245 | | |
| Conversion→Diocletian | 31 | 276 | 5776 |
| Diocletian→Council of Nicaea, year 12 of Constantine | 59 | 335 | 5835 |
| Nicaea→Constantinople | 58 | 393 | 5893 = 11C + 41 |
| Constantinople→Ephesus | 55 | [448] | [5948] |
| Ephesus→Chalcedon | [21] | 469 | 596[9] |
| Chalcedon→Islam | 170 | 639 | 6139 |

continued with Ethiopic rulers to Yoḥanes (J 1695); cf. **57**

## 13 1: *BM 827* $120^a$ I, 16—$120^b$ I, 9

| creation of Adam→Incarnation: 5500 | years | total |
|---|---|---|
| Adam→birth of Noah | 1656 | |
| Noah→recession of Flood | 600 | 2256 = 4C + 128 |
| recession of Flood→Tower | 571 | 2827 = 5C + 167 |
| Tower→Abraham | 501 | 3328 = 6C + 136 |
| Abraham→Moses | 425 | 3753 = 7C + 29 |
| birth of Moses→David | 694 | 4447 = 8C + 191 |
| David→Nābukadanaṣor | 469 | 4916 = 9C + 128 |
| Nābukadanaṣor→'Eskender | 265 | 5181 |
| 'Eskender Maqēdonāwi→birth of Christ | 319 | 5500 = 10C + 180 |

## **13** 2: *BM 827* $120^{b}$ I, 9—II, 3

| Christ | |
|---|---|
| conception | Magābit (VII) 29, Sunday |
| birth | Tāḫśāś (IV) 29, Tuesday, reign of Augustus |
| baptism | Ṭer (V) 11, Tuesday 5531, Tiberius year 16 |
| crucifixion | Magābit (VII) 27 |
| resurrection | Magābit (VII) 29 |
| ascension | Genbot (IX) 8 |

| Additional data | | EAC p. 59—61 |
|---|---|---|
| birth | t = 1, e = 9 | k = 181, c = 10 |
| baptism | t = 3, e = 11 | 211 2 |
| crucifixion | e = 14, m = 16 | 214 5 |

## **13** 3: *BM 827* $120^b$ II, 4—$121^a$ I, 3

| | | | |
|---|---|---|---|
| Augustus lived after birth of Christ | | | 14 years |
| Tiberius | ruled | | 23 |
| Gaius | | | 4 |
| Claudius | | | 14 |
| Matthew wrote Gospel in year | Claudius | 1 | = after ascension 8 years |
| Mark | Claudius | 4 | 11 |
| Luke | Claudius last year | | 22 |
| Nero | ruled | | 13 |
| John wrote Gospel in year | Nero | 8 | |
| execution of Peter and Paul | Nero | 13 | "on the 5th of Hamlē (XI), on a Sunday" |
| Vespasian | ruled | | 9 |
| Destruction of Temple | Vespasian | 6 | = after ascension 40 years = W 5574 |

## **134**: *BM 827* $121^{a}$ I, 3—$121^{b}$ II, 1

| | | years | J | total | |
|---|---|---|---|---|---|
| 1. | Birth of Christ→Conversion of Ethiopia | 245 | | | 1. |
| | Conversion→Diocletian | 31 | 276 | 5776 | |
| | Diocletian→Council of Nicaea, year 12 of Constantine | 59 | 335 | 5835 | |
| | Nicaea→Constantinople | 58 | 393 | 5893 = 11 C + 41 | |
| 5. | Constantinople→Ephesus | 55 | [448] | [5948] | 5. |
| | Ephesus→Chalcedon | [21] | 469 | 596[9] | |
| | Chalcedon→Islam | 170 | 639 | 6139 | |

continued with list of Ethiopic kings until Joḥanes (A. D. 1681) and Takla Hāymānot (A. D. 1706); cf. **57** 2

### **14:** *BM Add 16217* 13$^{a}$ I,3—13

| | | |
|---|---|---|
| Augustus lived after birth of Christ | | 15 years |
| Tiberius | ruled | 23 |
| Gaius, h. s. | | 4 |
| Neron | | 13 |
| three rulers | | 2 |
| Vespasian | ruled | 9 |
| Destruction of Temple | Vespasian 6 | = J 77 = after ascension 46$^{y}$ |

## **15:** *BM Add 16217* $19^{a}$ I, 2—$19^{b}$ I, 4

| | years | total | |
|---|---|---|---|
| (Adam)→Noah | 1656 | | |
| birth of Noah→Flood | 600 | 2256 | |
| Flood→Tower | 571 | 2827 | |
| Tower→Abraham | 520[!] | 3328 | |
| Abraham→Moses | 125[!] | 3753 | |
| birth of Moses→David | 699[!] | 4447 | |
| David→Nābukadanaḍor | 469 | 4916 | |
| Nābukadanaḍor→Alexander | 265 | 5181 | |
| Alexander Macedon→birth of Christ | 319 | 5500 | |
| birth of Christ→Diocletian | 277[!] | 5776 | Σ |
| Diocletian→Nicaea | 41 | | 5817 |
| Nicaea→Constantinople | [56] | | 5873 |
| [Constantinople]→Ephesus | 50 | | 5923 |
| Ephesus→Chalcedon | 21 | 168 | 5944 |
| birth of Christ→Chalcedon | 444 | | 5944 |
| Chalcedon→Islam | 170 | | 6114 |
| Adam→Islam | | 2114[!] | |
| birth of Christ→Islam | 614 | | 6114 |
| Diocletian→Islam | 338 | | 6114 |
| Islam[!]→Yekueno Amlāk | | 6762 | |

continued with Ethiopic rulers until ʾIyāsu (W 7223—7247); cf. **58**

**16:** *BM Add 16217* $20^{a}$ I, 4—II, 11

| | | text | margin | |
|---|---|---|---|---|
| Adam | lived | $950^{y}$ | 930 | |
| Sēt | | 917 | 902 | |
| | | | Hēnos | 905 |
| Qāynān | | 920 | 910 | |
| Malāle'ēl | | 805 | 1095 | |
| Yārēd | | 972 | 962 | |
| Hēnok, his time | | — | | |
| Lāmēk | | 776 | Lāmēh | 746 |
| Mātusālā | | 999 | | |
| Noḫ | | 950 | | |
| Sēm | | 700 | 500 | |
| 'Alfāksad | | 465 | 440 | |
| Qāhānem | | 430 | 440 | |
| ʿĒbor | | 434 | Sālā | 430 |
| Fālēq | | 430 | | |
| Rāgew | | 230 | 230 | |
| Sēruḥ | | 236 | 207 | |
| 'Aḫzab (~Nāmrud) | | 69 | | |
| Adam→Nāmrud | 3000 | | | |
| Nākor | | | Nākor | 129 |
| Tārā | | | Tārā | 205 |
| Sārā | | | Sārā | 127 |
| 'Abrehām | | | 'Abrehām | 175 |
| Yesḥaq, Yāʿeqob, 12 tribes | | | | |

## 17: *BM Add 16217* 20[a] II, 11—21[a] I, 12

| | | |
|---|---|---|
| subject to Pharaoh | | 430[y] |
| subject to Moses | | 40 |
| Faneḥas | | 25 |
| Kuesa, revolting | | 8 |
| Gotolyāl, s. o. Nēqez, brother of Kāleb | | 50 |
| ʿĒglom, king of Moʿāb | | 18 |
| Nāʿod, s. o. Gērā | | 80 |
| Nāʿod 26 | W 4000 | |
| Sēmēgēr | | 25 |
| ʾIyāmin | | 25 |
| Diborā and Bārq | | — |
| the Midians | | 7 |
| Gēdiwon | | 44 |
| ʾAbēmēlēk | | 3 |
| Tolā, s. o. Fuhā | | 23 |
| ʾIyāʾēr, the Gileadite | | 22 |
| the Philistines | | 18 |
| Yoftāḥē | | 7 |
| Ḥasēbon | | 10 |
| Sēlom, from Zābelon | | 10 |
| Lābon | | 8 |
| Philistines | | 40 |
| Somson, tribe of Mikā | | 20 |
| no ruler | | 12 |
| ʾĒli, the priest | | 40 |
| Sāmuʾēl, the prophet | | 22 |

## **18**: *BM Add 16217* $21^{a}$ I, 12—II, 18

| | |
|---|---|
| Sā'ol, first king | $40^{y}$ |
| Dāwit | 40 |
| Salomon | 40 |
| Robe'ām, h. s. | 17 |
| 'Iyorbe'am, h. s. | — |
| 'Abyā, h. s. | $7^{y}69^{d}$ |
| 'Asef | 41 |
| 'Iyosafeṭ | 25 |
| 'Iyorām | 8 |
| 'Akāzyās | 1 |
| Gotolyā | 8 |
| 'Ayosyās | 40 |
| 'Amēsyās | 30 |
| | Σ 297 |

## **19** 1: *BM Add 24995* $30^{b}$ II, 1—$31^{a}$ I, 4

| | years | total | |
|---|---|---|---|
| Adam→Noah | 1656 | | |
| birth of Noah→Flood | 600 | 2256 | |
| Flood→Tower | 571 | 2827 | |
| Tower→Abraham | 501 | 3328 | |
| Abraham→Moses | 425 | 3753 | |
| Moses→David | 694 | 4447 | |
| David→Nābukadanaḍor | 469 | 4916 | |
| Nābukadanaḍor→Alexander | 265 | 5181 | |
| Alexander→birth of Christ | 319 | 5500 | 180 in cycle 11 |
| Christ→Conversion of Ethiopia | 245 | 5745 | |
| Conversion→Nicaea | [90] | [5835] | |
| Nicaea→Gabra Masqal, s. o. Kālēb | 94 | 5929 | 77 in cycle 12 |

continued for Ethiopic rulers to Fasiladas (W 6781); cf. **60**

## **19**2: *BMA 24995* $32^{a}$ I, 9—25

| | years | Σ ↓ | Σ ↑ |
|---|---|---|---|
| 12 C = Gabra Masqal | 6384 | | |
| 13 C | 6916 | | |
| Adam→Flood | 2108 | | 1611 |
| Flood→Construction of the Tower | 558 | 2666 | 2169 |
| Construction of the Tower→Abraham | 1500 | 4166 | 3669 |
| Abraham→Exodus of Moses | 440 | 4606 | 4109 |
| Exodus of Moses→Alexander of Macedon | 1137 | 5743 | 5246 |
| Alexander→Cleopatra the Egyptian | 247 | 5990 | 5493 |
| Cleopatra the Egyptian→birth of Christ | 43 | 6033 | 5536 |
| Adam→birth of Christ, total | 5536 | | |

followed by a detailed chronology of Christ until Ascension in 5569: $32^{a}$ I, 25—$32^{b}$ I, 7

## **20:** *BN 160 A:* $2^a$, 1—11

| The Persians | | | |
|---|---|---|---|
| after capture Israel remained in Persia 70$^y$: | | | |
| Nābukadanaṣor kept them 26 | | | Nebuchadnezzar |
| Sērēyālmādāroq reigned 23 | | | Evil Merodach |
| Belṭāsor (and) Sēyēlmādāroq reigned 21 | | Σ | Neriglissar and Nabonid |
| Dāreyos, s. o. 'Aḥesurs of Māhi, and Kuerš of Persia ruled | 9$^y$ | 9 | Darius I and Cyrus |
| of which both of them (ruled) 7$^y$: | | | |
| after the death of Dāreyos Qiros reigned alone | 3 | 12 | |
| Faḥsayos s. o. Qiros reigned | 8 | 20 | Cambyses? s. o. Cyrus |
| Dāreyos s. o. Yesaṣef reigned | 3 | 23 | |
| Dāreyos Masaglāy reigned | 13 | 36 | Darius I |
| 'Asasurs h. s. reigned | 20 | 56 | Ahasuerus (= Xerxes I) |
| 'Azdārēs reigned | 40 | 96 | Artaxerxes I |
| Na?'edāsēr, the second reigned | 5 | 101 | Xerxes II |
| Ṣā'erinos reigned | 3 | 104 | Sogdianos |
| Dāreyos s. o. 'Amat, called Mantu reigned | 16 | 120 | Darius II |
| 'Azdeyāsēr h. s., brother of Qiros reigned | 14 | 134 | Artaxerxes II |
| 'Abdāsēr, called 'Akuš reigned | 20 | 154 | Artaxerxes III Ochus |
| 'Arsēs h. s. reigned | 4 | 158 | Arses |
| Dārā s. o. 'Arses reigned | 20$^y$6$^m$ | 178½ | Darius III |
| total | 276$^y$ | | |

## **21**: *BN 160 C*: $2^{a}$, 20—28: *Ptolemies*

| Ptolemaios (baṭlimos) | years | | years |
|---|---|---|---|
| Claudius(?) (qleyādiqos) | 7 | Philip Arrhidaeus | 7 |
| Alexander, his brother | 24 | Alexander II | 12 |
| Argob | 24 | Ptol. Soter | 20 |
| Philadelphos (mafqarē 'eḫuhu) | 16 | Philadelphos | 38 |
| Euergetes (ta'azāzi) | 5 | Euergetes | 25 |
| Philopator (m. 'abuhu) | 17 | Philopator | 19 |
| Epiphanes (banṣuḥ) | 24 | Epiphanes | 24 |
| Philopator (m. 'abuhu) | 25 | Philometor | 35 |
| Euergetes(?) (zagaber) | 20 | Euergetes | 29 |
| Soter (madeḫen) | 18 | Soter | 36 |
| Philometor (m. 'emu) Physkon (fasqos) | 10 | | |
| Auletes (qasa'os) | 18 | | |
| Dionysos (yenāsayos) | 24 | Dionysos | 29 |
| Cleopatra | 30 | Cleopatra | 21 |
| | [Σ: 262] | total | 293 |

## **22**: *BN 160 D*: $2^{a}$, 28—40: *Romans*

| | | |
|---|---|---|
| Augustus Caesar, s. o. Manuhos | (ruled) | $52^{y}6^{m}$ |
| birth of Christ | | |
| in the year 25 of his reign as Caesar; from Adam | | 5500 |
| Kanun (IV) 25, lunar Ṭebet (X) 10 | | |
| Šaban (VIII) 10 of the lunar year 5569 | | |
| Herod, regnal year 34 | | |
| Tiberius Augustus | ruled | 23 |
| crucifixion: | | |
| in his 18th regnal year | | |
| lunar Nisan (VIII) 15, Magābit (VII) 27, Ḥedar 25 | | |
| [resurrection:] | | |
| Delqa'edā (XI) 17, of the lunar year 5563 | | |
| Gaius | | |
| Claudius | ruled | 14 |
| Nero | | 13 |
| Galba (gābyos) | | $9^{m}$ |
| Otho ('aynun) | | $3^{m}$ |
| Vitellius (faṭolos) | | $3^{m}$ |

## **23**: *BN 160* $16^{a}$ I, 13—II, 18 and *BN 160* $80^{b}$ I, 17—$81^{a}$ I, 5

| | years | Σ |
|---|---|---|
| Adam→Flood | 2150 | |
| Flood→Abraham leaving Chaldea | 1199 | 3349 |
| Abraham leaving Chaldea→Exodus | 430 | 3779 |
| Exodus→end of Judges | 505 | 4284 |
| end of Judges, Kings→capture of Jerusalem | $516^{y}7^{m}4^{d}3^{h}$ | 4800 |
| Captivity→return from Captivity | 70 | 4870 |
| return from Exile→birth of Christ | $434 = 7^{w}$ of $62^{y}$ | 5304 |

| | $16^{a}$ II, 15—18 | | $81^{a}$ I, 1—5 | |
|---|---|---|---|---|
| no Judges or Kings | 172 | 5476 | 225 | 5529 |
| total | 5496 | | 5500 | |

## 24: *BN 160* 76ᵃ I, 1—13

"We wrote (this) Computus from Adam to the present day."

| | years | Σ |
|---|---|---|
| Adam→Flood | 6257 | 2257 |
| Flood→birth of Abraham | 1072 | 3329 |
| birth of Abraham→Exodus | 507 | 3836 |
| total | 3836 | |

*BN 160* 76ᵃ I, 14—76ᵇ II, 8

"Concerning the Judges"

| | | |
|---|---|---|
| Moses, s. o. 'Enbārm, first judge. In the wilderness | | 40ʸ |
| 'Iyāsu, s. o. Nawē, tribe of 'Ēfrēm | ruled | 34 |
| Kuesa 'Artēmu, king of Syria, revolting | | 8 |
| Gotonyāl, s. o. Q̣ēnēz | ruled | 50 |
| 'Ēglom, king of Mo'āb | | 18 |
| Nā'od, s. o. Gērā, tribe of 'Ēfrēm | | 80 |
| Śēmēgār, s. o. Ḥanāt, killed 600 (Philistines) | | — |
| 'Iyāmin | | 25 |
| Bārq and Diborā | | 40 |
| Gēdeyon, s. o. Yo'ās, tribe of 'Ēfrēm | | 40 |
| 'Abamālēk, h. s. | | 3 |
| Tolā Foḥa, tribe of Yesākor | | 23 |
| 'Iyā'ar, the Gileadite | | 22 |
| Yoftāḥē, the Gileadite | | [6] |
| Ḫāsēbon, of Bethlehēm | | 7 |
| 'Ēli'um, from 'Ēfrātā | | 8 |
| Philistines | | 40 |
| Samson, s. o. Menāḥē, tribe of Dān | | 20 |
| 'Ēli, the priest, tribe of Lēwi | | 20 |
| Samu'ēl, the priest, tribe of Lēwi | | 20 |

## **25:** *BN 160* 76$^{b}$ II, 8—77$^{a}$ II, 5
## "Concerning the kings of Judah" — "From Israel"

| | |
|---|---|
| Sā'ol, s. o. Qisa, first king, tribe of Benyām | $40^{y}$ |
| Dāwit, s. o. 'Esēy, tribe of Yehudā | 40 |
| Salomon, h. s. | 40 |
| Robe'ām, h. s. | 17 |
| 'Abyā, h. s. | $6^{y}8^{d}3^{h}$ |
| 'Asā, h. s. | 44 |
| Yosāfeṭ, h. s. | 25 |
| 'Iyorām | 8 |
| 'Akazyās | 10 |
| Gotonyāl, m. o. 'Akazyās | 10 |
| Yo'ās, s. o. 'Akazyās | 40 |
| 'Amisyās | 29 |
| 'Azāryās and 'Ozyān | 52 |
| 'Iyo'atām | 16 |
| 'Ākaz | 17 |
| Ḥezeqyās | 29 |
| Menāsē | 52 |
| 'Āmoṣ | $2^{y}12^{d}$ |
| 'Iyokes | 34 |
| 'Iyo'akaz | $3^{m}$ |
| 'Iyo'iqim | $14^{y}14^{d}$ |
| 'Iyo'aqim, h. s., 'Ikonyās | $3^{m}$ |
| Mātān, h. s., Sēdēqyās | 14 |
| kings of Yehudā | $526^{y}7^{m}3^{d}$ |

## 26: *BN 160* 77[a] II, 5—77[b] I, 19; l. marg.

| [Kings of Israel] | years |
|---|---|
| ʾIyorbeʿām s. o. Nābāṭ, ruled | 24 |
| Nābāṭ, h. s. | 2 |
| Baʾas | 24 |
| ʾĒla, h. s. | 2 |
| Zenberi | 12 |
| ʾAkaʾab, h. s. | 22 |
| ʾAkāzyās, h. s. | 2 |
| ʾIyorām, his brother | 12 |
| killed ʾIyu and ruled | 28 |
| ʾIyoʾakaz, h. s. | 17 |
| ʾIyorbeʾām, h. s. | 41 |
| ʾAzāryās, h. s. | 6[m] |
| Sēlom, h. s. | 30[d] |
| Menāḥē | 10 |
| Fāqēsyās | 2 |
| Fāquḥē, h. s. | 28 |
| Hosēʾe | 9 |
| total, 18 (kings) | 257[y]6[m] |
| **Kings of Edom** | |
| Bādāq, s. o. Bēʿor; ʾIyobāb, s. o. Zārā; ʾAdād, s. o. Bārād; ʾAsmā; Saʾol, s. o. Robat; Balaʾinon | no dates |
| **Kings of Samaria** | |
| ruled until Ḥezeqyās | 290 |
| captivity in Persia | 125 |
| tribe of Judah in Persia | 70 |
| Nāka | |

## 27: *BN 160* 77ᵇ II, 1 — 78ᵇ II, 2

| | | | | |
|---|---|---|---|---|
| Adam had lived | 230ʸ | | | when Sēt was born |
| Sēt | 205 | hence total W | 430 | Hēnos |
| Hēnos | 190 | | 620 | Qaynān |
| Qāynān | [100] | | 725 | Malale'ēl |
| Malāle'ēl | 165 | | 890 | Yārod |
| Yārod | 162 | | 1052 | Hēnok |
| Hēnok | 162 | | 1214 | Mātusālā |
| Mātusālā | 167 | | 1384 | Lamēḫ |
| Lamēḫ | 128 | | 1569 | Noḫ |
| Noḫ | 500 | | 2069 | Sēm |
| Sēm | 100 | | 2169 | 'Arfāksed |
| 'Arfāksed | 135 | | 2304 | Qāynān |
| Qāynān | 130 | | 2434 | Sālā |
| Salā | 130 | | 2564 | 'Ēbēr |
| 'Ēbēr | 130 | | 2694 | Fālēq |
| Fālēq | 130 | | 2824 | Rāgew |
| Rāgew | 132 | | 2956 | Sēroḫ |
| Sēroḫ | 135 | | 3091 | Nākor |
| Nākor | 109 | | 3200 | Tārā |
| Tārā | 100 | | 3300 | 'Abrehām |
| 'Abrehām | 60 | | 3360 | leaving Chaldea |
| people of Israel | 430 | | 3790 | in Kanā'an and Egypt |

## **28:** *BN 160* $78^b$ II, 2—$79^a$ II, 11

| | | |
|---|---|---|
| | | 3790 |
| Moses ruled | $40^y$ | |
| 'Iyasus | 31 | hence total: W 3861 |
| Kuesa, revolting | 8 | |
| Gotonyāl, s. o. Qēnēz | 5 | |
| 'Ēglom, king of Mo'ab | 18 | 3892 |
| Nā'od | 80 | |
| 'Iyāmēn | 25 | 3907 |
| Sēmēgār | — | |
| Diborā and Bārq | 40 | |
| the Midians | 7 | 3954 |
| Gēdēwon | 40 | |
| 'Abamēlēk, h. s. | 3 | |
| Tolā, s. o. Foḥā | 23 | 4020 |
| 'Iyā'ar, the Gileadite | 22 | |
| Philistines | 18 | |
| Yoftāḥē | 6 | 4066 |
| Ḥasēbon | 7 | |
| 'Ilomu, from Zābelon | 10 | |
| Lobon, from 'Afrāta | 8 | 4091 |
| Philistines | 40 | |
| Somson | 20 | 4151 |
| without Judges | 12 | |
| 'Ēli, the priest | 20 | |
| Samu'ēl | 22 | 4205 |

## **29**: *BN 160* $79^{a}$ II, 12—$80^{a}$ I, 9

| | | |
|---|---|---|
| | | 4205 |
| Sā'ol, first king, tribe of Benyām, ruled | $40^{y}$ | |
| Dāwit, tribe of Yehudā | 40 | |
| Salomon, h. s. | 40 | hence total: W 4325 |
| Robeʿām, h. s. | 17 | |
| 'Abyā, h. s. | $6^{y}8^{d}[3]^{h}$ | |
| 'Asā, h. s. | 41 | 4389 |
| 'Iyosāfeṭ, h. s. | 25 | |
| 'Iyorām, h. s. | 8 | |
| 'Akazyās, h. s. | 1 | |
| Gotolyā, m. o. 'Akazyās | 6 | 4409 |
| 'Iyosā, s. o. 'Akazyās | 40 | |
| 'Amēsyās, h. s. | 29 | |
| 'Azāryās | 52 | 4550 |
| 'Iyo'atām, h. s. | 16 | |
| 'Ākāz, h. s. | 16 | |
| Ḥezeqyās | 29 | |
| Menāsē | 55 | |
| 'Ameṣ, h. s. | $2^{y}12^{d}$ | |
| 'Iyosayās, h. s. | 31 | 4699 |
| 'Iyo'akāz | $3^{m}$ | |
| 'Iyo'āqēm→'Ikonyān | $11^{y}14^{d}$ | |
| 'Iyo'āqim | $3^{m}$ | |
| Mātān→Sēdēqyās | 11 | 4730 |
| until capture of Jerusalem | $516^{y}7^{m}4^{d}3^{h}$ | |
| captivity | 70 | 4810 |
| return from captivity→Christ | $434^{y} = 7^{w}$ of $62^{y}$ | 5500 |

## **30**: *EMML 215* $64^a$ II, 23—$64^b$ I, 28

| "Computus for 13 cycles" | |
|---|---|
| Cycle 1 | Adam→Ḫēnos 97 |
| Cycle 2 | Ḫēnos 98→[Yārēd] 104 |
| Cycle 3 | Yārēd 105→Lāmēh 122 |
| Cycle 4 | Lāmēh 123→Noḫ 472 |
| Cycle 5 | [Noḫ 473]→'Ebēr 128 |
| Cycle 6 | 'Ebēr 129→Nākor 4 |
| Cycle 7 | Nākor 5→Amram 42 |
| Cycle 8 | [Amram 43]→Gidēwon, the judge, 1 |
| Cycle 9 | Gidēwon 2→Ḥezeqyās 22 and $6^m$ |
| Cycle 10 | Ḥezeqyās 23½→Ptolemy 15 |
| Cycle 11 | Ptolemy 16→Martyrs 76 |
| Cycle 12 | Martyrs 77→Martyrs 608 |
| Cycle 13 | Martyrs 609→Martyrs 1140 |
| | total: 6917 1 2 3 4 5 6 7 |

## **31**: *EMML 215* $72^b$ I, 21—$73^a$ II, 7

| | years | total | | n. C | n |
|---|---|---|---|---|---|
| Adam→birth of Noah | 1656 | | − 60 = 3C | 1596 | 3 |
| Noah→Flood | 600 | 2256 | − 128 4 | 2128 | 4 |
| Flood→Tower | 571 | 2827 | − 167 5 | 2660 | 5 |
| Tower→Abraham | 501 | 3328 | − 136 6 | 3192 | 6 |
| Abraham→Moses | 425 | 3753 | − 29 7 | 3724 | 7 |
| Moses→David | 694 | 4447 | − 191 8 | 4256 | 8 |
| David→Nābukadanaṣor | 469 | 4916 | − 128 9 | 4788 | 9 |
| Nābukadanaṣor→Alexander | 265 | 5181 | − 393 9 | | |
| Alexander→Christ | 319 | 5500 | − 180 10 | 5320 | 10 |
| Christ→Conversion of Ethiopia | 245 | 5745 | − 425 10 | | |
| Conversion→Diocletian | 31 | 5776 | | | |
| Diocletian→Nicaea | 59 | 5835 | − 515 10 | | |
| Nicaea→Gabra Masqal | 94 | 5929 | − 77 11 | 5852 | 11 |
| Gabra Masqal→ Yekuno 'Amlāk | 377 | 6300 | | | |

## **32:** *EMML 215* 73$^{a}$ II, 15—73$^{b}$ II, 8

| | years | total | |
|---|---|---|---|
| Adam→birth of Noah | 1656 | | −60=3C |
| birth of Noah→Flood | 600 | 2256 | |
| Flood→Tower | 571 | 2827 | |
| Tower→Abraham | 501 | 3328 | |
| Abraham→Moses | 425 | 3753 | |
| Moses→Dawid | 694 | 4447 | |
| Dawid→Nābukadanaṣor | 469 | 4916 | |
| Nābukadanaṣor→Alexander | 265 | 5181 | |
| Alexander→birth of Christ | 319 | 5500 | |
| Christ→conversion of Ethiopia | 245 | 5745 | |
| conversion of Ethiopia→Diocletian | 31 | 5776 | |
| Diocletian→Nicaea | 59 | 5835 | |
| conversion of Ethiopia→Nicaea | 90 | | |
| Nicaea→Gabra Masqal | 94 | 5929 | |

followed by list of Ethiopic kings; cf. **67**

## **33:** *EMML 215* 73$^{b}$ II, 27—74$^{a}$ I, 23

| | years | total | Σ | [Σ] |
|---|---|---|---|---|
| Adam→Noah | 2032 | | | 2000 |
| Noah→Isaak, s. o. Abraham | 1604 | | 3636 | 3600 |
| Isaak→Moses | 430 | | 4066 | 4030 |
| Moses→building of Jerusalem | 430 | | 4496 | 4460 |
| building of Jerusalem→Ezra | 540 | 5000 | 5036 | 5000 |
| Ezra→birth of Christ | 500 | | 5536 | 5500 |
| birth of Christ→Abreha and Aṣbeḥa | 235 | | 5781 | 5745 |
| Abreha and Aṣbeḥa→ Gabra Masqal, s. o. Kālēb | 182 | | 5963 | 5927 |
| Gabra Masqal→Zāguē | 240 | | 6203 | 6167 |
| Zāguē→Yekuno Amlāk | 133 | 6372 | 6336 | 6300 |

## **34**: *EMML 2063* $26^a$ II, 11—$27^a$ I, 8

| | years | Σ | [Σ] |
|---|---|---|---|
| Adam→Flood | 2128 | | |
| Flood→Tower | 540 | 2668 | |
| Tower→Abraham $75^y$ | 1071 | 3739 | |
| Abraham $76^y$→Exodus | 430 | 4169 | |
| Exodus→Temple | 440 | 4609 | |
| Temple→Captivity | 324 | 4933 | 4930 |
| Captivity→Ezra | 70 | 5003 | 5000 |
| Ezra→birth of Christ | 530 | 5533 | 5536 |
| birth of Christ→year 1 of Martyrs | 316 | 5849 | 5852 |
| year 1 of Martyrs→spread of Christianity | 114 | 5963 | 5963 |
| spread of Christianity→year 1 of Gabra Masqal | 421 | 6384 | =12C |
| year 1 of Gabra Masqal→completion of 13 cycles | 532 | 6916 | |
| additional $84^y$ to 13C: total | 7000 | | |

## **35**: *EMLL 2063* $27^a$ I, 9—II, 16

| | | years |
|---|---|---|
| Adam→birth of Christ | | 5536 |
| Adam→ascension | | 5569 |
| Adam→Diocletian 1 | | 5852 |
| Diocletian with Maximianus, ruled | $28^y$ | |
| Constantine ruled | 40 | |
| Constantine 10: Council of Nicaea | | |
| Adam→Ṣaḥam 28 | | 6177 |
| ʾela Amidā | 5 | |

## **36**: *EMML 2063* $27^{b}$ I, 1—II, 15

| "Number of years according to the Computus of the Copts, from our father Adam to the coming of our Lord, Jesus Christ, praise to him" | years | Σ |
|---|---|---|
| Adam→Noah | 1642 | |
| Noah→Flood | 600 | 2242 |
| Flood→construction of the Tower | 350 | 2592 |
| constr. of the Tower→dispersion of languages | 208 | 2800 |
| dispersion of lang.→Lord speaking to Abraham | 240 | 3040 |
| our father Abraham→Exodus | 840 | 3880 |
| from then→Alexander | 1093 | 4973 |
| Alexander→Cleopatra | 294 | 5267 |
| from her→annunciation by Gabriel | 233 | 5500 |
| total | 5500 | |
| Incarnation→years of the Martyrs | 276 | 5776 |

## **37**: *EMML 2063* $27^{b}$ II, 16—$28^{a}$ II, 12

| "And from another book" | years | Σ |
|---|---|---|
| Adam→Flood | 2256 | |
| Flood→construction of Tower | 578 | 2834 |
| construction of Tower→Lord speaking to Abraham | 569 | 3403 |
| our father Abraham→Exodus | 432 | 3835 |
| Exodus→reign of David | 606 | 4441 |
| David→destruction of Jerusalem | 488 | 4929 |
| Jerusalem→reign of Alexander | 263 | 5192 |
| Alexander→end of reign of Cleopatra | 279 | 5471 |
| from it→birth of Christ | 30 | 5501 |
| Christ→Diocletian | 276 | 5777 |
| total | 5776 | |

## 38: *EMML 2063* 44[a] II,5—44[b] II,16

| | |
|---|---|
| "Computus of the cycles" | |
| Cycle 1 | Adam→'Ēnos 97 |
| Cycle 2 | 'Ēnos 98→Yārēd 104 |
| Cycle 3 | Yāred 105→Lāmēḫ 122 |
| Cycle 4 | Lāmēḫ 123→Noḫ 472 |
| Cycle 5 | Noḫ 473→'Ebēr 128 |
| Cycle 6 | 'Ebēr→Nākor 4 |
| Cycle 7 | Nākor 5→'Amram 42 |
| Cycle 8 | 'Amram 43→Gēdewon, the judge, 1 |
| Cycle 9 | Gēdewon 2→Ḥezeqyās $22^{y}6^{m}$ |
| Cycle 10 | Ḥezeqyās 23½→Ptolemy 15 |
| Cycle 11 | Ptolemy 16→Martyrs 76 |
| Cycle 12 | Martyrs 77→Martyrs 608 |
| Cycle 13 | Martyrs 609→Martyrs 1140 |
| | total: 6916 |

## 39: *EMML 2063* 46[a] II,7—47[a] II,2

| | years | total | |
|---|---|---|---|
| Adam→Flood | | 2256 | |
| Flood→Tower | 544 | 2800 | |
| Tower→appear. of Lord to Abraham | 640 | 3440 | |
| appear. of Lord to Abraham→ Moses the prophet | 440 | 3880 | |
| Moses→[birth of Christ] | 1620 | 5500 | Σ |
| birth of Christ→years of Martyrs (era D) | 276 | | 5776 |
| years of Martyrs→council of Nicaea | 59 | | 5835 |
| Nicaea→council of Constantinople | 58 | 117 | 5893 |
| Constantinople→council of Ephesus | 55 | 172 | 5948 |
| Ephesus→council of Chalcedon | 21 | 193 | 5969 |
| Chalcedon→Islam | 140 | 333 | 6109 |
| beginning of World→Islam | | 6109 | |
| | 891 | 7000 | |

## **40:** *EMML 2063* 47$^{a}$ II,2—47$^{b}$ I,8

| | years | total | |
|---|---|---|---|
| Adam→Noah | 1656 | | |
| Noah→Flood | 600 | 2256 | |
| Flood→Tower | 571 | 2827 | |
| Tower→birth of Abraham | 501 | 3328 | |
| Abraham→birth of Moses | 425 | 3753 | |
| Moses→David | 694 | 4447 | |
| David→Nābukadanaṣor | 469 | 4916 | |
| Nābukadanaṣor→Alexander | 265 | 5181 | |
| Alexander→birth of Christ | 319 | 5500 | |
| Christ→Diocletian | 276 | 5776 | Σ |
| Diocletian→Nicaea | 41 | | 5817 |
| Nicaea→Constantinople | 56 | | 5873 |
| [Constantinople]→Ephesus | 50 | | 5923 |
| Ephesus→Chalcedon | 21 | 168 | 5944 |
| Christ→Chalcedon | 444 | | |
| Chalcedon→Islam | 170 | | 6114 |
| Adam→Islam | | 6114 | |
| Islam→Yekueno 'Amlāk | 648 | | 6762 |
| Adam→Yekueno 'Amlāk | | 6762 | |

continued with Ethiopic rulers until Yāʿeqob 1 (J 1598); cf. **68**

## **41:** *EMML 2077* 154b I, 17—155a I, 6

| | | | | | | | | | | |
|---|---|---|---|---|---|---|---|---|---|---|
| 154b I, 17 | ’Adām | created | Sēt | at | 230y | and lived thereafter | 700y | hence total lifetime | 930y | |
| | Sēt | | Hēnos | | 205 | | 707 | | 972 | |
| | Hēnos | | Qāynān | | 190 | | 715 | | 905 | |
| | Qāynān | | Malāle’ēl | | 170 | | 740 | | 910 | |
| | Malāle’ēl | | Yārēd | | 165 | | 640 | | 805 | |
| | Yārēd | | Hēnoẖ | | 162 | | 800 | | 962 | |
| | Hēnoẖ | | Mātusālā | | 165 | | 200 | | 365 | removed by the Lord |
| | Mātusālā | | Lāmēẖ | | 187 | | 782 | | 969 | |
| | Lāmēẖ | | Noẖ | | 182 | | 595 | | 777 | |
| | hence ’Adām→ | | Noẖ | | 1656 | | | | | |
| 154b II, 24 | Noẖ | created | Sēm | at | 500 | and lived after the Flood | 350 | total lifetime | 950 | |
| | Sēm | | ’Arfāskad | | 100 | and lived thereafter | 500 | and died | | |
| | ’Arfāskad | | Qāynām! | | 135 | | 335 | and died | | |
| | Qāynān | | Sālā | | 130 | | 440 | | | |
| | Sālā | | ’Ēbēr | | 130 | | 430 | | | |
| | ’Ēbēr | | Fālēq | | 134 | | 430 | | | |
| | Fālēq | | Rāgew | | 130 | | 270 | | | |
| | Rāgew | | Sēroh | | 232 | | 269 | | | |
| | Sēroh | | Nākor | | 136 | | 200 | | | |
| | Nākor | | Tārā | | 79 | | 129 | | | |
| | Tārā | | ’Abrehām | | 70 | | 250 | | | |
| | hence Noẖ→ | | ’Abrehām | | 1776 | | | | | |
| | ’Adām [→ | | ’Abrehām] | | 3432 | | | | | |
| 154b III, 28 | ’Abrehām | created | Yesḥaq | at | 100 | and lived thereafter | 75 | | | |
| | Yesḥaq | | Yāʿeqob | | 60 | | | | | |
| | Yāʿeqob | [ ] | Fāreʿon | | 130 | | | | | |
| | stay in Egypt and Kana’an | | | | 430 | | | | | |
| | hence ’Abrehām→ | | Exodus | | 720 | | | | | |
| | ’Adām [→ | | Exodus] | | 4152 | | | | | |

## 42: *EMML 2077* 155$^{a}$ I, 7—23

| | | |
|---|---|---|
| [ | —] | |
| Kuesā | $8^{y}$ | |
| Gotoniʿāl | 50 | |
| ʾĒglon | 18 | |
| Nāʾod | 80 | |
| Midians | 7 | |
| Gidēwon | 40 | |
| Diborā | 40 | |
| Tolā | 23 | |
| ʾIyāʾēr | 22 | |
| Philistines | 18 | |
| Yoftāḥē | 6 | |
| Hēsēbol | 7 | |
| ʾĒlom | 10 | |
| Lobon | 8 | |
| Philistines | 40 | |
| Somson | 20 | |
| Exodus→Judges | | $397^{y}$ |
| from Adam | | 3829 |

## **43:** *EMML 2077* $155^{a}$ I, 23—II, 13

| | | | |
|---|---|---|---|
| 'Ēli | ruled | $20^{y}$ | |
| Dāwit | reigned | 40 | |
| Salomon | | 40 | |
| Robe'ab | | 17 | |
| ʿAbiyu | | 3 | |
| 'Asā | | 41 | |
| 'Iyosāfeṭ | | 25 | |
| 'Iyorām | | 8 | |
| 'Akāzyās | | 4 | |
| Gotolyā | | 7 | |
| 'Iyo'as | | 40 | |
| 'Amēsyās | | 29 | |
| 'Azāryās | | 52 | |
| 'Iyo'atām | | 16 | |
| Ḥezeqyās | | 29 | |
| Menāsē | | 55 | |
| 'Amoṣ | | 2 | |
| Yoseyās | | 81 | |
| 'Ēliyāqim | | 11 | |
| 'Iyo'aqim | | $3^{m}$ | |
| Nātānyān | | 21 | Σ: $541^{y}3^{m}$ |
| Judges→Kings→Captivity | | | $616^{y}3^{m}$ |
| from Adam | | | 4445 |
| to Ezra | | | 5500 |

After a short section on dispersal of languages continued with Ethiopic kinglist (until Takla Hāymānot); cf. **69**

## 44: *ME* Guidi [1896] p. 379 I,23—380 I,18

| lunar | | solar | | event |
|---|---|---|---|---|
| year | day | year | date | |
| 2311 | 12 | 2256 | IX 26 Sun | Flood |
| 3960 | 14 | 3844 | VII 23 Wedn | Exodus |
| | | 3844 | 24 Thrs | Exodus |
| | 6 | 3844 | VIII 15 Thrs | Exodus |
| | 6 | 3844 | IX 15 Sat | Exodus |
| | | 4982 | VIII 30 Tue | Temple (Cyrus) |
| | | 5042 | IV 26 | Temple (Ezra) |
| | | 5067 | I 16 Sun | Temple (Artaxerxes) |
| | | 5182 | II 4 Mo | Alexander, beg. of reign |
| | | 5484 | III 17 Thrs | Augustus, beg. of reign |
| 5667 | 1 | 5500 | VII 29 Sun | Christ, conception |
| 5668 | 9 | 5501 | IV 29 Tue | Christ, birth |
| 5697 | 22 | 5531 | V 11 Tue | Christ, baptism |
| 5700 | 15 | 5534 | VII 27 Fri | Christ, crucifixion |
| | 17 | 5534 | 29 Sun | Christ, resurrection |

## 45: *Princeton 5884* 7$^{b}$ III,10—20

| | | |
|---|---|---|
| Augustus | lived | 15$^{y}$ after the birth of Christ |
| Tiberius | ruled | 23$^{y}$ |
| Gaius, h. s. | | 4 |
| Claudius, h. s. | | 14 |
| Nero | | 13 |
| three rulers | | 2 |
| Vespasianus | | 9 |
| in his 6th year: destruction of the Temple in the 77th year from the birth of Christ | | |

## **46:** *Princeton 5884* $12^b$ II, 5—$13^a$ I, 14

| | years | total | |
|---|---|---|---|
| Adam→birth of Noah | 1656 | | |
| Noah→recession of Flood | 600 | 2256 | |
| recession of Flood→building of Tower | 571 | 2827 | |
| building of Tower→Abraham | 501 | 3328 | |
| Abraham→Moses | 425 | 3753 | |
| Moses→David | 703 | 4456 | |
| David→Nābukadanaṣor | 460 | 4916 | |
| Nābukadanaṣor→Alexander | 265 | 5181 | |
| Alexander→birth of Christ | 319 | 5500 | |
| Christ→Diocletian | 276 | 5776 | |
| Diocletian→Nicaea | 41 | | 5817 |
| Nicaea→Constantinople | 56 | | 5873 |
| Constantinople→Ephesus | 50 | | 5923 |
| Ephesus→Chalcedon | 21 | 168 | 5944 |
| Christ→Chalcedon | | 444 | 5944 |
| Chalcedon→Islam | 170 | | |
| Adam→Islam | | 6114 | |
| Christ→Islam | 614 | | |
| Diocletian→Islam | 338 | | |
| Islam→Yekuno 'Amlāk | 648 | | |
| Adam→Yekuno 'Amlāk | | 6762 | |

continued for Ethiopic rulers to Zadengel (A. D. 1604), with later additions to 1790; cf. **70**

## **47:** *Upps 3* $63^a$, 2—$63^b$, 3

| | | W | |
|---|---|---|---|
| Adam→Noah | 2000 | | |
| Noah→Isaak, s. o. Abraham | 1600 | 3600 | |
| Isaac→Moses | 430 | 4030 | |
| Moses→building of Jerusalem | 430 | 4460 | |
| building of Jerusalem→Ezra | 540 | 5000 | |
| Ezra→birth of Christ | 500 | 5500 | 180 in cycle 11 |

continued for Ethiopic rulers to Zare'e Yā'eqob (W 6471); cf. **71**

## 48: *Vat 1* $205^a$ II, 41 — $205^b$ II, 11

| from | | to | | Cycles |
|---|---|---|---|---|
| Adam | year 1 | Hēnos | year 97 | 1 C = $532^y$ |
| Hēnos | 98 | Yārid | 104 | 2 |
| Yārid | 105 | Lāmēḫ | 142 | 3 |
| Lāmēḫ | 143 | Noḫ | 86 | 4 |
| Noḫ | 87 | Ebēr | 22 | 5 |
| Ebēr | 23 | Nākor | 28 | 6 |
| Nākor | 29 | Abrehām | 64 | 7 |
| Abrehām | 65 | A'lom | 6 | 8 |
| Agālom | 7 | Yonātān, king of Yehudā | 7 | 9 |
| Yonātān, king of Yehudā | 8 | Ṭalomiyos Faluda lada | 36 | 10 |
| Ṭalomiyos | 37 | Diocletian | 1 | 11 |
| Diocletian | 2 | Diocletian | 533 | 12 |
| Diocletian | 534 | Diocletian | 1065 | 13 |
| from Adam to Diocletian 1065: $6916^y$ | | | | |

## **49:** *Vat 1* 205ᵇ II, 28—206ᵃ I, 20

| | | | | total |
|---|---|---|---|---|
| Adam | → | removal of Enoch at 365ʸ | 1455ʸ | |
| removal of Enoḫ | → | Flood | 787 | 2242 |
| Flood | → | Tower | 558 | 2800 |
| Tower | → | Abraham at 75ʸ | 588 | 3388 |
| Abraham at 75ʸ | → | birth of Isaac | 25 | 3413 |
| Isaac at 60ʸ | | birth of Jacob | 60 | 3473 |
| Jacob 2000ʸ | | birth of Lēwi | [83] | [3556] |

## **50:** *Vat 1* 207ᵃ I, 17—II, 1

| Adam, year 1 | end | beg. |
|---|---|---|
| 'Ēnos | 97 | 98 |
| 'Iyārēd | 104 | 105 |
| Lāmēk | 112 | 113 |
| Noḫ | 86 | 87 |
| 'Ebār | 22 | 23 |
| Nākor | 28 | 29 |
| 'Abrehām | 64 | — |
| 'Egālom | 35 | 36 |
| Yonātān, king of Yehudā | — | 38 |
| Paṭolamiyos Mafqarē 'Eḫuhu | 3[7]? | 38 |
| Diocletian year 1 | — | — |

## **51:** *Vat 1* 207ᵃ II, 25—39

| | years | Σ |
|---|---|---|
| Adam→Flood | 2068 | |
| Flood→75th year of Abraham | 1256 | 3324 |
| 75th year of Abraham→Exodus | 430 | 3754 |
| Exodus→construction of the Temple | 440 | 4194 |
| Construction of the Temple→Captivity | 934 | 5128 |
| Captivity→birth of Christ | [434] | 5562 |

continued for Ethiopic rulers to Ṣaḥam 28 (W 6177); cf. **72**

## 52: *Vat 1* $207^{b}$ I, 16—36

| | | years |
|---|---|---|
| Adam→birth of Christ | | 5562 |
| Adam→ascension | | 5594 |
| Adam→Diocletian I | | 5860 |
| Diocletian with Maximianus, h. s., ruled $28^{y}$ | | |
| Constantine | 40 | |
| Constantine 10: Council of Nicaea | | |
| Adam→Ḍeḥam 28 | | 6177 |
| 'ela Amidā | 5 | |

## 53: *Vat 1* $204^{a}$ II,39—$205^{a}$ I,8

| | years | total |
|---|---|---|
| Adam→birth of Lēwi | 3556 | |
| birth of Lēwi→birth of Moses | 180 | 3736 |
| Exodus at Moses $80^{y}$ | 80 | 3816 |
| in the desert | 40 | 3856 |
| after the desert→Tusa 26 | 57 | 3913 |
| until judge Zatune'ēl | 45 | [3]958 |
| until judge Gēdeyon | 203 | 4161 |
| until judge Yeftāḥēl 6 | 69 | 4230 |
| until Samson judged $20^{y}$ | 85 | 4315 |
| until Sāmu'ēl judged $20^{y}$ | 85 | 4395 |
| until death of David | [100] | 4495 |
| death of David→Olympiad 1,1 | 250 | 4745 |
| until death of Alexander | 428 | 5173 |
| death of Alexander→death of Cleopatra | 284 | 5457 |
| death of Cleopatra→birth of Christ | [43] | 5500 |
| until crucifixion | 33 | 5533 |
| resurrection→birth$^{!}$ of Diocletian | 243 | 5776 |
| birth$^{!}$ of Diocletian→D 76 | 76 | 5852 = 11 C |

## **54**: *Vindob 6* $2^{b}$, 17—26

| | years | Σ |
|---|---|---|
| Creation→Flood | 2256 | |
| Flood→Exodus | 1588 | 3844 |
| Exodus→reign of Saul | 539 | 4383 |
| reign of Saul→Captivity in Babylon | 512 | 4895 |
| Captivity in Babylon→return | 70 | 4965 |
| return→Alexander | 206 | 5171 |
| Alexander→Augustus | 277 | 5448 |
| Augustus→birth of Christ | 40 | 5488 |
| birth of Christ→Islam | 614 | 6102 |
| Islam→Takla Hāymānot | 591 | 6693 |

# Ethiopic Kings

## **55**: *Berol 84* 22[b] II, 12—23[a] I, 26; continued from **7**

| | | years | Σ |
|---|---|---|---|
| Council of Constantinople→Gabra Masqal, s. o. Kāleb, king of Ethiopia | | 36 | 5929 |
| Gabra Masqal→Zāguay | | 244 | 6173 |
| Zāguay→Yekueno 'Amlāk | | 133 | 6306 |
| Yekueno 'Amlāk | ruled | 15 | |
| Yāgeb'a Ṣeyon, h. s. | ruled | 9 | |
| 2 of his sons | ruled | 3 | |
| 3 of his sons | ruled | 2 | |
| Wedem Ra'ād, h. s. | ruled | 15 | |
| 'Āmda Ṣeyon, s. o. Wedem Ra'ād | ruled | 30 | |
| Yekueno 'Amlāk→Sayf 'Ar'āda | | 74 | 6380 = 12·532—4 |
| year 4 of Sayfa 'Ar'ād: completion of 12 cycles | | | |
| year 5 of Sayfa 'Ar'ād: beginning of cycle 13 | | | |
| Sayfa 'Ar'ād | ruled | 28 | |
| Wedem 'Asfarē, h. s. | ruled | 10 | |
| Dāwit with Tēwodros, h. s. | | 32 | |
| Sayfa 'Ar'ād→Yesḥaq, s. o. Dāwit | | 70 | 6450 |
| Yesḥaq with 'Endreyās, h. s. | | 16 | |
| Ḥezba Nān with 2 of his sons | ruled | 5 | |
| Yesḥaq year 1→Zar'a Yā'eqob, s. o. king Dāwit | | 21 | 6471 |
| Zar'a | ruled | 34 | |
| Ba'eda Maryām, h. s. | ruled | 10 | |
| 'Eskender, s. o. Ba'eda Maryām with h. s. 'Āmda Ṣeyon | ruled | 16 | |
| Nā'od | ruled | 13 | |
| Zar'a Yā'eqob year 1→Lebna Dengel | | 73 | 6544[=7000—456] |
| Lebna Dengel | ruled | 32 | 6576[=7032—456] |

**56**: *BM 754* 4[a] I, 9—II, 11; continued from **8**

| | | years | Σ |
|---|---|---|---|
| Islam→Yekueno 'Amlāk | | 648 | 6762 |
| Yekueno 'Amlāk | ruled | 15 | 6777 |
| Yāgebā Ṣeyon, h. s. | | 9 | 6786 |
| other 5 sons: Bāḥr, Sagad, Senf Sagad | | | |
| Žan Sagad, Ḥezba Rāʿād, Qedem Sagad | | 5 | 6791 |
| Wedem Re'ed | | 15 | 6806 |
| ʿĀmda Seyon | | 30 | 6836 |
| Sayf ʿĀr'ad | | 28 | 6864 |
| Wedem 'Asfarē | | 10 | 6874 |
| Dāwit with Tēwodros, h. s. | | 32 | 6906 |
| Yesḥaq with 'End(re)yās, h. s. | | 16 | 6922 |
| Ḥezb Nāñ, ʿĀmda 'Iyasus, Bādel Nāñ | | 5 | 6927 |
| Zar'a Yāʿeqob | | 34 | 6961 |
| Ba'eda Māryām | | 10 | 6971 |
| 'Eskender | | 16 | 6987 |
| Nāʿod | | 13 | 7000 |
| Wanāg Sagad | | 32 | 7032 |
| 'Asnāf Sagad | | 19 | 7051 |
| 'Admās Sagad | | 9 | 7060 |
| Malak Sagad | | 39 | 7099 |
| Yāʿeqob, h. s., with Zadengel | and he reached | 10 | |

| | years | total | Σ |
|---|---|---|---|
| Islam → Yekuno 'Amlāk | 622 | | |
| from birth of Christ | 1262 | 6762 = 12C + 378 | |
| Yekuno 'Amlāk ruled | 15 | | 6777 |
| Yāgeb'a Ṣeyon | 9 | | 6786 |
| 5 of his sons: Bāḥr 'Asgad, Ṣenf Sagad, Žan Sagad, Ḥezb 'Arʿād, Wadem Sagad | 5 | | 6791 |
| Wadem 'Arʿād | 15 | | 6806 |
| ʿĀmda Ṣeyon, h. s. | 30 | | 6836 |
| Sayf 'Arʿād, h. s. | 28 | | 6864 |
| Wedem 'Asfarēm, h. s. | 10 | | 6874 |
| Dāwit, h. s. | 29 | | 6903 |
| Tēwodros | 3 | | 6906 |
| Yesḥaq with 'Endreyās | 16 | | 6922 |
| in year 10 of his rule: completion of 13 C | | 6916 = 13C | |
| from Yekuno 'Amlāk to that time | 154 | | |
| Ḥezb Nāñ with 2 of his sons: ʿĀmda 'Iyasus and Badel Nāñ | 5 | | 6927 |
| Zar'a Yāeqob | 34 | | 6961 |
| Ba'eda Māryām, h. s. | 10 | | 6971 |
| 'Eskender, h. s. | 16 | | 6987 |
| Nā'od, h. br. | 13 | | 7000 |
| from year 11 of Yesḥaq to the death of Nā'od | 84 | 7000 | |
| Lebna Dengel, h. s. | 32 | | 7032 |
| Galāwdewos, h. s. | 19 | | 7051 |
| Minās, h. br. | 4 | | 7055 |
| Śarḍa Dengel, h. s. | 34 | | 7089 |
| Yāʿeqob | 7 | | 7096 |
| Zadengel, h. s. | 1 | | 7097 |
| without king | 1 | | 7098 |
| Yāʿeqob, again | 1 | | 7099 |
| Susneyos, throne name Śelṭān Sagad | $25^{y}6^{m}9^{d}$ | | |
| Fāsiladas, throne name ʿĀlam Sagad | $35^{y}1^{m}$ | | |
| Yoḥanes, h. s., throne name 'A'elāf Sagad | $14^{y}9^{m}7^{d}$ | | 7174? |
| 'Iyāsu, h. s., throne name 'Adyām Sagad | $24^{y}$ | | |

**57** 2: *BM 827* $121^{b}$ II, 1—$122^{a}$ II, 11; continued from **13** 4

| | years | total | Σ |
|---|---|---|---|
| Islam→Yekuno 'Amlāk | 622 | | |
| from birth of Christ | 1262 | 6762 = 12C + 378 | |
| Yekuno 'Amlāk | 15 | | 6777 |
| Yāgeb'a Ṣeyon | 9 | | 6786 |
| 5 of his sons: Bāḥr 'Asgad, Senf Sagad, | | | |
| Žan Sagad, Ḥezb 'Arʿād, Wedem Sagad | 5 | | 6791 |
| Wedem 'Arʿād | 15 | | 6806 |
| ʿĀmda Ṣeyon, h. s. | 30 | | 6836 |
| Sayfa 'Arʿād, h. s. | 28 | | 6864 |
| Wedem 'Asfarē[m], h. s. | 10 | | 6874 |
| Dāwit, h. s. | 29 | | 6903 |
| Tēwodros | 3 | | 6906 |
| Yesḥag with 'Endreyās, h. s. | 16 | | 6922 |
| in year 10 of his rule: completion of 13C | | 6916 = 13C | |
| from Yekuno 'Amlāk to that time | 154 | | |
| Ḥezb Nāñ with 2 of his sons: ʿĀmda 'Iyasus and Badel Nāñ | 5 | | 6927 |
| Zar'a Yāeqob | 34 | | 6961 |
| Ba'eda Māryām, h. s. | 10 | | 6971 |
| 'Eskender, h. s. | 16 | | 6987 |
| Nā'od, h. br. | 13 | | 7000 |
| from year 11 of Yesḥaq to the death of Nā'od | 84 | 7000 | |
| Lebna Dengel, h. s. | 32 | | 7032 |
| Galāwdewos, h. s. | 19 | | 7051 |
| Minās, h. br. | 4 | | 7055 |
| Śarḍa Dengel, h. s. | 34 | | 7089 |
| Yāʿeqob | 7 | | 7096 |
| Zadengel, h. s. | 1 | | 7097 |
| without king | 1 | | 7098 |
| Yāʿeqob again | 1 | | 7099 |
| Susneyos, throne name Śelṭān Sagad | $25^{y}6^{m}9^{d}$ | | |
| Fāsiladas, throne name ʿĀlam Sagad | $35^{y}1^{m}$ | | |
| Yoḥanes, h. s. | $14^{y}9^{m}7^{d}$ | | $7174^{?}$ |
| 'Iyāsu, h. s., throne name 'Advām Sagad | $24^{y}7^{m}20^{d}$ | | |

## **58**: *BM Add 16217* 19^b^ I, 2—20^a^ I, 4; continued from **15**

| | years | total | Σ = W |
|---|---|---|---|
| Islam→Yekueno 'Amlāk | | 6662 | 6762 |
| Yekueno 'Amlāk ruled | 15 | | 6777 |
| ʿĀmda Ṣeyon, h. s. | 30 | | 6836 |
| Sayfa 'Ar'ed, h. s. | 28 | | 6864 |
| Wedema 'Asfar, h. s. | 10 | 112 | 6874 |
| Dāwit | 29 | | 6903 |
| Tēwodros, h. s. | 3 | | 6906 |
| Yesḥaq with 'Endreyās | 17 | | 6923 |
| Ḥezba Nāyn with his sons | 5 | | 6928 |
| Zar'a Yāʿeqob | 49 | | 6962 |
| Ba'eda Māryām, h. s. | 10 | | 6972 |
| 'Eskender, h. s. | 17 | | 6988 |
| Nā'od | 13 | 227 | 7001 |
| Wanāg Sagad | 34 | | 7035 |
| 'Aṣnāf Sagad | 19 | | 7054 |
| 'Admās Sagad | 4 | | 7058 |
| Yāʿeqob, h. s. | 9 | | 7067 |
| Zadengel | 1 | | 7068 |
| Śelṭan Sagad | 26 | | 7094 |
| ʿĀlam Sagad, h. s. | 35 | | 7129 |
| Yoḥanes, h. s. | 15 | | 7144 |
| 'Iyāsu | 24 | | 7168 |
| Taklahāymānot, h. s. | $2^{y}6^{m}$ | | $7171^{?}$ |
| Tēwoflos | 4 | | 7175 |
| Yosṭos | 3 | | 7178 |
| Dāwit | 5 | | 7183 |
| Bakāfā | $9^{y}3^{m}15^{d}$ | | $7192^{?}$ |
| 'Iyāsu | $24^{y}9^{m}25^{d}$ | | $7217^{?}$ |
| 'Iyoʿās | $40^{y}$ | | 7257 |

## **59**: *BMA 16217* 21^b^ I, 1—17

| | |
|---|---|
| [ ] and Asew→Ye[k?uno A?mlāk | . . . kin]gs |
| Arwē → [Bāzēn] | 25 kings |
| [Bāzēn]→Arbeḥa and Aṣ[be]ḥa | 19 (kings) |
| Barḥa and Aṣbeḥa→Gabra Masqal | 25 years |
| Gabra Masqal→Anbas Wedem | 20 years |
| Naʿod ḍeposed [ ] thrones with Zaguē receivi?ng [Yekuno A]mlāk | |
| And [Takla] Hāymānot | [. . year]s |
| Yekuno A[mlā]k→Aṣē Yoḥanes | 84 years |

**60**: *BM Add 24995* 31ª I, 1—II, 17; continued from **19 1**

| | | years | total | Σ |
|---|---|---|---|---|
| Nicaea→Gabra Masqal, s .o. Kālēb | | 94 | 5929 = 11C + 77 | |
| Gabra Masqal→Zāguay | | 244 | 6173 321 | |
| Zāguay→Yekuno Amlāk | | 133 | 6306 454 | |
| Yekuno Amlāk→Sayfa Raʿād | | 74 | 6380 528 | |
| year 4 of Sayfa Raʿād: completion of 12 C | | | | 6384 |
| Sayfa Raʿād→Zarʿā Yāʿeqob | | 91 | 6471 = [12 C + ] 87 | |
| Zarʿā Yāʿeqob | ruled | 34 | | |
| Baʿeda Māryā!, h. s. | | 10 | | |
| ʾEskender | | 16 | | |
| Nāʿod | | 13 | | |
| Lebna Dengel | | 32 | | |
| Galāwdēwos | | 19 | | |
| Minās | | 4 | | |
| Śarḍa Dengel | | 34 | | |
| Yāʿeqob | | 9 | | |
| Zadengel | | 1 | | |
| until Susenyos year 16 | | 15 | + 187 | |
| Susenyos, called Śelṭān Sagad | ruled | $25^{y}7^{m}8^{d}$ | | |
| Adam→Fāsiladas, h. s. | | | 6781 | |
| according to Abušākers computus | | | 7138 | |
| $13\,C + 84^{y}$ | | | 7000 | |

**61: *BN 160***

| | *A*: $7^b$, 1—20 | | *A′*: $90^a$ II, 6—14 | | | |
|---|---|---|---|---|---|---|
| | Years of the kings of Aksum | | kings | | | |
| 1. | Arwē | 400 | 1 | Arwē | (?) | 1. |
| | Engābo killed Arwē | | | Engābo expelled and killed Arwē | $200^y$ | |
| | Gedur | 200 | 2 | Gedur in Nuḥ | 50 | |
| | with Nabu'e | 100 | 3 | | | |
| 5. | Sabaṣ, in Sad and in Aksum | 50 | 4 | Sabaḍa in Sāda 50 and in Aksum | 50 | 5. |
| | discussion? with? Mikidā | 50 | 5 | | | |
| | because of Terad with Solomon | | | because of Terad, Jerusalem | | |
| | Queen of the South visits Solomon | | | | | |
| | in year 4 of Solomon, year 36 of Sa'ol | | | | | |
| | after that she ruled 25 years | | | | | |
| 11. | Arwē→Mākidā | 5 kings [800] | | | | 11. |
| 12. | Bayna Leḥkam, s. o. Solomon, ruled | 29 | 1 | | | 12. |
| | Ḫedādyo | 4 | 2 | | | |
| | Awd | 11 | 3 | | | |
| 3 | Wāsayo | 3 | 4 | | | |
| 16. | Ṣaw'e | 44 | 5 | | | 16. |
| | | Σ: [91] | | | | |
| 17. | Qāsyo→Qatr of Māweṭ | 64 | | | | 17. |
| | Baḥas | 9 | 1 | | | |
| | Qāwdā | 2 | 2 | | | |
| 20. | Qanaz | 10 | 3 | | | 20. |
| | Hedunā | 9 | 4 | | | |
| | Wazḫā | 4 | 5 | | | |
| | Hedunā II | 6 | 6 | | | |

| | A: 7^b, 1—20 | | A′: 90^a II, 6—14 | |
|---|---|---|---|---|
| | Years of the kings of Aksum | | kings | |
| | Safāyā | 16 | 7 | |
| 25. | Zefēlyā | 27 | 8 | 25. |
| | Amālub | 3 | 9 | |
| | Birwās | 29 | 10 | |
| | Maḥasi Balḥawā | 17 | 11 | |
| | Serotu | 27 | 12 | |
| 30. | Bākes | 10 | 13 | 30. |
| | Masinh | 6 | 14 | |
| | Satwā | 9 | 15 | |
| 33. | Adgalā | $10^{y}6^{m}$ | 16 | |
| | Amābā'e | $6^{m}$ | 17 | |
| 35. | Malisu | $4^{y}$ | 18 | 35. |
| | Ḥaqāle | 13 | 19 | |
| | Dedemēhē | 10 | 20 | |
| | Weṭeṭ | 2 | 21 | |
| | Aldā | 30 | 22 | |
| 40. | Zagen and Rēmā | 8 | 23 | 40. |
| | Gefālē | 4 | 24 | |
| | Śarq | 4 | 25 | |
| | Asguāgua | 6 | 26 | |
| | Hesqa | 21 | 27 | |
| 45. | Ṣawāza | $1^{m}$ | 28 | 45. |
| | Wākena | $2^{d}$ | 29 | |
| | Adawe | $9^{m}$ | 30 | |
| | Sagal | $3^{y}$ | 31 | |
| 49. | Nālkē | (?) | 32 | 49. |
| | | Σ: [301 + 64] | | |
| 50. | Bāzēn year 8: birth of Christ | | | 50. |
| | Leḥkem→Bāzēn 8: 45 kings | $984^{y}$ | | |
| | [illegible] | $1800^{y}$ | | |

## 62: *BN 160*

| | *B*: $7^{b}$, 20—25 | | *C*: $7^{b}$, 25—30 | | |
|---|---|---|---|---|---|
| 1. | za'ela Asfeḥa | $14^{y}$ | za'ela Abreha | 10 | 1. |
| | za'ela Ṣagab | 23 | za'ela Asfeḥa | 3 | |
| | 'ela Śamrā | 3 | za'ela Śāhl | 14 | |
| | za'ela Yebā'e | 17 | za'ela Reteʿe | 1 | |
| 5. | za'ela Eskendē | 37 | za'ela Ēsfeḥ | 5 | 5. |
| | za'ela Ṣaḥam | ? | za'ela Aṣbeḥa | 16 | |
| | za'ela San | 39 | za Aminādā | 7 | |
| | za'ela Eyegā | 13 | za Abrāḫ | $2^{m}$ | |
| | za'ela Aminādā | $18^{y}1^{m}$ | za Śāhl | $2^{m}$ | |
| 10. | za'ela Aḫyāwā | 3 | za Gabaz | 14 | 10. |
| | za'ela Aṣbeḥa and Abreha | $26^{y}7^{m}$ | za Sehul | 4 | |
| | year 4 of their reign: conversion of Aksum | | za Aṣbeḥa | 3 | |
| | Birth of Christ→conversion: 13 kings | $245^{y}$ | za'ela Abreḫa za'ela Eder | 17 | |
| | | | za'ela Ṣaḥam | 18 | |
| | | | za Amidāb | 12 | 15. |
| | | | za Śāhl | 2 | |
| | | | za Asbāḥ | 15 | |
| | | | za'ela Zebaz | 14 | |
| | | | za Agālē waza Lēwi | 2 | |
| | | | za Aminādāb | 11 | 20. |
| | | | za Yāʿeqob waza Dāwit | 3 | |
| | | | za Armāḫ | $14^{y}7^{m}$ | |
| | | | za Zitānā | 12 | |
| | | | za Yāʿeqob | 9 | |
| | | | za Qēsṭantinos | 28 | 25. |
| | | | | Σ: $235^{y}1^{m}$ | |

## **63**: *BN 160*

### *D:* $7^{b}$, 30—36

| Years of the kings of Aksum: | | |
|---|---|---|
| 'Arwē→Nālkē | 51 kings | [1]859$^{y}$ |
| Bāzēn | ruled | 17$^{y}$ |
| Bāzen year 8: birth of Christ | | |
| Bāzen→'Abreha and 'Aṣbeḥa, y. 13: 18 kings | | 245 |
| 'Arwē→'Aṣbeḥa and 'Abreha | 85 kings | 2104 |
| from the construction of the Temple: 426$^{y}$ from 11 C | | |

### *E:* $7^{b}$, 36—$8^{a}$, 21

| From the birth of Christ in the reign of Bāzēn: | kings |
|---|---|
| Ṣenfa 'Asgad | 1 |
| K. Nag.: Bāḥra 'Asgad | 2 |
| Germā | 3 |
| 'Asfar | 4 |
| Ser'ātā | 5 |
| Kuelu Laṣeyon | 6 |
| Śardāy | 7 |
| Zar'ay | 8 |
| Bagāmay | 9 |
| Ğan 'Asgad | 10 |
| Ṣeyon | 11 |
| Hagaz | 12 |
| Mālāy | 13 |
| Sayfa | 14 |
| 'Ar'ād | 15 |
| 'Agdār | 16 |
| 'Abreha and 'Aṣbeḥa, loving brothers | 17, 18 |
| 18 kings, 245$^{y}$ | |

## **64:** *BN 160*

*F:* $8^{a}$, 21—28

| | kings |
|---|---|
| K. Nag.: 'Asfeḥa | 1 |
| 'Arfed and 'Amša, brothers | 2, 3 |
| K. Nag.: 'Arʿādā | 4 |
| Sel'adbā | 5 |
| 'Alʿāmidā | 6 |
| Tāzēn | 7 |
| Kālēb | 8 |
| Gabra Masqal | 9 |
| From 'Abreha and 'Aṣbeḥa→Gabra Masqal: | 9 kings $184^{y}$ |

| | kings |
|---|---|
| Qēsṭanṭinos | 1 |
| Zezegār | 2 |
| 'Asfaḥ | 3 |
| 'Armāḥ | 4 |
| 'Arzān | 5 |
| 'Asfaḥ | 6 |
| Ğān 'Asgad | 7 |
| Ferē | 8 |
| Śanay | 9 |
| 'Adr'az | 10 |
| 'Ayzār | 11 |
| Delnaʿad Māʿedāy | 12 |
| ʿEṣit. s. o. ʿEsut. Amhara, years of disorder | |
| 'Aubas Wedem | 13 |
| Kala Wedem | |
| Germā | 14 |
| 'Asfar | 15 |
| Ğer Gaʿāz | 16 |
| Degnā | 17 |
| Mikā'el | 18 |
| Bada Gaʿāz | 19 |
| 'Armāḥ | 20 |
| Ḥazba Nāñid | 21 |
| Ğuanāğan | 22 |
| K. Nag.: 'Anbasā Wedem | 23 |
| Delna'ad | 24 |
| From Gabra Masqal→Delna'ad: 24 kings $244^{y}$ | |

**65**: *BN 160 G:* 8ª, 29—36

| | | | kings |
|---|---|---|---|
| Transfer of kingship to a nation not from the tribe of David | | | |
| From Zague→Yekuno 'Amlāk | | 133ʸ | |
| Restoration of the kingship to Yekuno 'Amlāk | | | 1 |
| 'Agbe'a Ṣeyon | | | 2 |
| Bāḥr Sagada | | | 3 |
| Ḥezba 'Arʿed | | | 4 |
| Qedma Sagad | | | 5 |
| Ǧān 'Asgad | | | 6 |
| Wedem Raʿād | | | 7 |
| 'Amda Ṣeyon | | | 8 |
| From Yekuno 'Amlāk→Sayfa 'Arʿāda | 8 kings | 74ʸ | |
| Sayfa 'Arʿāda | | | 1 |
| Wedem 'Asfar | | | 2 |
| Dāwit | | | 3 |
| Tēwodros | | | 4 |
| From Sayfa 'Arʿād→Yesḥaq | 4 kings | 70ʸ | |
| Adam→Yesḥaq | | 6400ʸ[=6856—456] | |

**66** 1: *BN 160* 16ª II, 19—16ᵇ II, 9

| | | years | total | Σ |
|---|---|---|---|---|
| Bāzēn→Abreha and Aṣbeḥa | | | [244] | |
| [A. and A.]→Gabra Masqal | 9 kings | 184 | 428 | |
| Gabra [M.]→Del Na'od | 20 kings | 235 | 663 | |
| kingship taken away, for | | 250 | 913 | 913 |
| thereafter kingship restored to Israel by the Lord | | | | |
| Yekueno Amlāk→Sayfa Arʿāda | 8 kings | 74 | 984 | 987 984 |
| Sayfa Arʿāda→year 1 of Zare'a Yāʿeqob | | 91 | — | 1078 1075 |
| reign of Zare'a Yāʿeqob | | 34 | 1108 | 1112 1109 |
| error | | 129 | 1239 | 1241 1238 |
| Ba'eda Māryām | | 10 | | 1251 |
| Eske'der | | 16 | | 1267 |
| Nā'od | | 13 | | 1280 |
| Lebna Dengel | | 33 | 1298 | 1313 1310 |
| Gālāwdēwos, time of writing | | — | | |

**66** 2: *BN 160* 80ª I, 9—80ᵇ I, 16

| In the days of Bāzēn, king of Ethiopia, and Augustus, king of Rome: birth of Christ | | years | total | Σ | W |
|---|---|---|---|---|---|
| [Christ]→Abreha and Aṣbeḥa | 19 kings | 244 | | | |
| Abreha and Aṣbeḥa→Gabra Masqal | 9 kings | 184 | 428 | | |
| Gabra Masqal→Del Nā'od | 20 kings | 235 | 663 | | |
| kingdom taken away, for | | 250 | 913 | | |
| restored by the Lord, remembering Dawid, through Yekueno Amlāk | | | | | |
| Yekueno Amlāk→Sayfa Arʿāda | 8 kings | 74 | 987 | 987 | 6487 |
| Sayfa Arʿāda→year 1 of Zare'a Yāeqob | | 91 | 1070 | 1078 | |
| reign of Zare'a Yāʿeqob | | 34 | 1108 | 1112 | |
| Bā'eda Māryām | | 10+... | | 1122+ | |
| Eskender | | 16 | | 1138 | |
| Nā'od | | 12 | | 1150 | |
| Lebna Dengel | | 32 | | 1182 | |
| writing of this book: year 13 of our king Gelāwdēwos, which is G | | 206 | | 1195 | 6695 |
| thereafter the earthqnake on Monday and Tuesday | | | 1192 | | |
| from Adam to this (time) | | | 6590 | | 6590 = 12C + 206 |

**67**: *EMML 215* 73ᵇ II, 3—27; continued from **32**

| | years | total |
|---|---|---|
| Nicaea→Gabra Masqal, s. o. Kāhen | 94 | 5929 |
| Gabra Masqal→Zāguē | 244 | 6173 |
| Zaguē→Yekuno Amlāk | 133 | 6306 |
| Yekuno Amlāk→Sayfa Areʿed | 74 | 6380 |
| Sayfa Areʿed→Yesḥaq | 70 | 6450 |
| Yesḥaq→Zareʿā Yāʿeqob, our king | 21 | 6471 |

**68**: *EMML 2063* 47ᵇ I, 7—II, 7; continued from **40**

| | years | total | Σ |
|---|---|---|---|
| Adam→Yekueno ʼAmlāk | | 6762 | |
| Yekueno ʼAmlāk ruled | 15 | | |
| Yāgbāʼa Ṣeyon, h. s. | 9 | | 6777 |
| other children: Baḥr ʼAsged, Senf ʼAsged, Žen ʼAsged, Ḥezb ʼArʿēd, Qedma ʼAsged | 5 | | 6786 |
| another son: Wadem Raʿād | 15 | | 6791 |
| ʿĀmda Ṣeyon | 30 | | 6806 |
| Sayfa ʼArʿād | 28 | | 6836 |
| Wedem | 10 | | 6864 |
| Dāwit with Tēwodros, h. s. | 32 | | 6874 |
| Yesḥaq with ʼEndreyās, h. s. | 16 | | 6906 |
| year 10 of Yesḥaq: completion of 13 cycles | | | 13 C = 6916 |
| Ḥezba Nāñ with 2 sons: ʿĀmda ʼIyasus, Badel Nāñ | 5 | | 6922 |
| Zarʼa Yāʿeqob | 34 | | 6927 |
| Baʼeda Māryām, h. s. | 10 | | 6961 |
| ʼEskender, h. s., with his son | 16 | | 6971 |
| Nāʿod, his brother | 13 | | 6987 |
| Lebna Dengel, h. s. | 32 | | 7000 |
| Galāwdēwos, h. s. | 19 | | 7032 |
| Minās, his brother | 9 | | 7051 |
| Śarḍa Dengel, h. s. | 39 | | 7060 |
| Yāʿeqob, h. s. | 9 | | 7099 |
| birth of Christ→year 1 of Yāʿeqob | | 1598 | |
| 1 lunar cycle: 532 years | | | |

## **69**: *EMML 2077* 155$^a$ II, 21—III, 31; continued from **43**

| | | | |
|---|---|---|---|
| Zague→Yekueno 'Amlāk | 336$^y$ | Wanāg Sagad Lebna Dengel | ruled 32$^y$ 4$^m$ 3$^d$ |
| Yekueno 'Amlāk ruled | 15$^y$ | 'Aṣnāf Sagad Galāwdewos | 15$^y$ 7$^m$22$^d$ |
| Yagbā Seyon | 9 | 'Admās Sagad Minas | 3$^y$10$^m$ |
| Žen Sagad | 6 | Malak Sagad Śarḍa Dengel | 34$^y$ 4$^m$19$^d$ |
| Qedem Sagad and Ḥezb 'Ar'ad | 3 | Yā'eqob | 9 |
| 3 sons of Yekueno 'Amlāk: | | Za Dengel | 1 |
| Zen 'Asgad, Ḥezb 'Asgad, Žen 'Ar'ad | 1 | Śelṭān Sagad Susenyos | 25 |
| Wedem 'Ar'ad | 15 | 'Ālam Sagad Fāsiladas | 35$^y$ 1$^m$ |
| 'Amda Ṣeyon 'Iwosṭatiwos (Eustatios) | 30$^y$10$^m$ | 'A'elāf Sagad Yoḥanes | 14$^y$ 9$^m$11$^d$ |
| Sayf 'Ar'ad, h. s. | 28$^y$ 2$^m$ | 'Adyām Sagad 'Iyāsu | 23$^y$ 7$^m$15$^d$ |
| Wedem 'Asfarē | 10 | Le'ul Sagad Takla Hāymānot | 2$^y$ 3$^m$20$^d$ |
| Dāwit | 32$^y$ 5$^m$ 9$^d$ | 'Aṣrār Sagad Tēwoflos | 3$^y$ 3$^m$ |
| Tēwodros | 9 | Bālgāda Yosṭos | 4$^y$ 1$^m$ |
| Yesḥaq | 15$^y$10$^m$ | Dāwit | 5$^y$ 3$^m$11$^d$ |
| °Edreyas | 7$^m$ | Bakāfā 'Iyāsu | [10$^y$]4$^m$ |
| Ḥezba Nāñ | 4$^y$ 2$^m$14$^d$ | [Berḥān Sagad] | 25 |
| 'Amda 'Iyasus | 3$^m$20$^d$ | 'Iyo'as, h. s. | 13$^y$10$^m$ |
| Baṣel Nāñ | 6$^m$13$^d$ | Yoḥanes with his son Takla Hāymānot | [ |
| Zare'a Yā'eqob | 30$^y$ 2$^m$ 8$^d$ | | |
| Ba'eda Maryām | 10$^y$ 2$^m$13$^d$ | 155$^b$: list of rulers and events in Ethiopic history from Menelik to Takla Hāymānot but without chronological data (mostly in Amharic) | |
| 'Eskender | 15$^y$ 6$^m$ | | |
| 'Amda Ṣeyon | 6$^m$ | | |
| Na'od | 13$^y$10$^m$ 8$^d$ | | |

## **70**: *Princeton 5884* 13[a] I, 10—II, 20; continued from **46**

| | | years | total | Σ |
|---|---|---|---|---|
| Islam→Yekuno 'Amlāk | | 648 | | |
| Adam→Yekuno 'Amlāk | | | 6762 | |
| Yekuno 'Am(lāk) | ruled | 15 | | 6777 |
| Yāgbā Ṣeyon, h. s. | | 9 | | 6786 |
| 5 of his sons | | 5 | | 6791 |
| Wedem Ra'ād | | 15 | | 6806 |
| 'Amda Seyon, h. s. | | 30 | | 6836 |
| Sayfa 'Ar'ạd, h. s. | | 28 | | 6864 |
| Wedem'asfārē, h. s. | | 10 | | 6874 |
| Dāwit | | 29 | | 6903 |
| Tēwodros, h. s. | | 3 | | 6906 |
| Yesḥaq and 'Endreyās, h. s. | | 17 | | 6923 |
| Ḥezb Nāñ with 2 of his sons | | 5 | | 6928 |
| Zar'a Yāēqob | | 34 | | 6962 |
| Ba'eda Māryām, h. s. | | 10 | | 6972 |
| 'Eskender, h. s. | | 17 | | 6989 |
| Nā'od | | 13 | | 7002 |
| Lebna Dengel, h. s. | | 32 | | 7034 |
| Galāwdēwos, h. s. | | 19 | | 7053 |
| Minās | | 4 | | 7057 |
| Śarḍa Dengel, h. s. | | 34 | | 7091 |
| Yā'ēqob, h. s. | | 9 | | 7100 |
| Zade(n)gel | | 1 | | 7101 |
| Susenyos | | 25 | | 7126 |
| Fāsil | | 35 | | 7161 |
| Yoḥanes | | 15 | | 7176 |
| 'Adyām Sagad 'Iyāsu | | 25 | | 7201 |
| Takla Hāymānot | | 2½ | | 7203½ |
| Tēwoflos | | 5 | | 7208 |
| Yosṭos | | 3 | | 7211 |
| Dāwit | | 5 | | 7216 |
| Bakāfā | | 9 | | 7225 |
| 'Iyāsu and 'Iyo'as | | 40 | | 7265 |
| Yoḥanes | | 1 | | 7266 |
| Takla Haymānot | | 8 | | 7274 |
| Takla Giyorgis | | 7 | | 7281 |
| from 'Ali to 'Ali | | 72 | | 7353 |
| Tēwodros | | 15 | | 7368 |
| Yoḥanes | | 22 | | 7390 |

**71:** *Upps 3* 63b, 3—64a, 16; continued from 47

| | years | W | |
|---|---|---|---|
| | | 5500 | 180 in cycle 11 |
| birth of Christ→Abreha and Aṣbeḥa | 245 | 5745 | 425 |
| Abreha and Aṣbeḥa→Gabra Masqal, s. o. Kālēb | 184 | 5929 | 77 in cycle 12 |
| Gabra Masqal→Zāguā | 244 | 6173 | 321 |
| Zāguā→Yekuno Amlāk | 133 | 6306 | 454 |
| Yekuno Amlāk→Sayfa Arʿāda | 74 | 6380 | 528; Sayfa 5 = cycle 13, year 1 |
| Sayfa Arʿāda→Zare'e Yāʿeqob | 91 | 6471 | 87 in cycle 13 |

## **72**: *Vat 1* 207$^{a}$ II,39—207$^{b}$ I,9; continued from **51**

| | | | | Σ |
|---|---|---|---|---|
| | | | | 5562 |
| [birth of Christ]→ | ’ela ’Abrehā, first year | | 473 | 6035 |
| and his reign? | ’ela ’Abrehā | 12 | | 6047 |
| | ’ela ’Afséḥa | 7 | | 6054 |
| | ’ela Śāhl | 14 | | 6068 |
| | ’ela ’Adẖano | 14 | | 6082 |
| | ’ela Retaʿe | 1 | | 6083 |
| | ’ela ’Asfeḥa | 5 | | 6088 |
| | ’ela ’Aṣboḥ | 10$^{?}$ ga$^{?}$ | | 6098 |
| | ’ela ʿAmidā | 7 | | 6105 |
| | ’ela Gabaz | 14 ga$^{?}$ | | 6119 |
| | ’ela Śehuʿul | 10$^{?}$ | | 6129 |
| | ’ela ’Aṣbuḥa$^{?}$ | 3 | | 6132 |
| | ’ela ’Abrehā | 17 | | 6149 |
| | ’ela Ḍeḥam | 28 | | 6177 |
| total from Adam to year 28, reign la’ela Ḍeḥam | | | 6177 | |

## Notes to the Texts

**1**: *B 84* $7^b$ II, 24—$8^a$ I, 21
Duplicate: **34**
Parallels: **7, 15, 40, 46; 19** 1; cf. Table 21 (p. 134) and Table 1 (p. 29)
l. 9: W 5852 = 11 C is the year 0 of the era G, not the year 1 of the era D (a common mistake; e. g. also in **2** line 3).
l. 10: "spread of Christianity"; waṣe'a krestenâ 'ag'âzi.
l. 11: W 6384 = 12C.

**2**: *B 84* $8^a$ I, 22—II, 7
Duplicates: **35, 52.**
l. 1—3 for variants in **52** against **2** and **35** cf. p. 43.
l. 7 and 8: cf. p. 65 and Table 22 (p. 145).

**3**: *B 84* $8^a$ II, 7—24
Duplicate: **36**
For "Coptic Computus" cf., e. g., *Vi 6* fol. $4^a$ where a 532-year table is called zagebṣâweyân.
l. 7: 208 restored from **36.**
l. 9: 438 error for 840; correct in **36.**
l. 11: Alexander→Cleopatra $294^y$: cf. **21** and Table 12 (p. 49).

**4**: *B 84* $8^a$ II, 24—$8^b$ I, 12
Duplicate: **37.**

**5**: *B 84* $18^b$ I, 1—II, 3
Duplicates: **30, 38.**
For related texts cf. Table 1 (p. 29) and Table 21 (p. 134).
According to these three texts, we have for the Flood, i. e. Noah 601: $4 \cdot 532 - 471 + 600 = 1657 + 600 = 2257$, a date which is also given in **24.**
l. 8, 9: Embarm = Amram, father of Moses (1 Chron. 6, 3).
l. 9: $22^y$ $6^m$: 2 Kings 18, 1, 2 give $29^y$.
l. 10, 11: Ptolemy: 'abṭalmêwos.
l. 15: I do not know what the numbers 1 to 7 mean. The same in **30** but not in **38.**

**6**: *B 86* $20^b$(19)—$21^a$ I, 28
Duplicate: **39;** Parallels: **12, 13** 4; **8.** Cf. Table 20 (p. 133).

**7:** *B 84* 22b I, 2—II, 16
Near duplicates: **15, 40, 46.**
Parallels: **1, 34; 19** 1; cf. Table 21 (p. 134) and Table 1 (p. 29).

**8:** *BM 754* 4a I, 1—9
Parallels: **12, 13** 4; **6, 39.** Cf. Table 20 (p. 133).

**9:** *BM 815* 17a I, 15—17b I, 3
Duplicate: **13** 1. For parallels cf. Table 1 (p. 29) and Table 21 (p. 134). A lengthy colophon (17a I, 1—15) names as author of the chronological sections (**13** 1 to **13** 4 and **57** 1) a well-known Coptic scholar, Girgis ibn Amid, who wrote in Damascus in the years 1262 to 1286[1]. For an English translation cf. Weld Blundell, Roy. Chron. p. 496—499.

**10:** *BM 815* 17b I, 3—19
Duplicate: **13** 2.
l. 4: 5531 is equated with Tiberius 16; therefore the crucifixion in 5534 would be Tiberius 19, not 18 as said in **22.**

All modern correspondences given by Weld Blundell (p. 497) for the Coptic dates are incorrect and should be as follows[2]:

| | Text | Julian |
|---|---|---|
| Conception | VII 29 [5500] | Mar 25 |
| Birth | IV 29 [5501] | Dec 25 |
| Baptism | V 11 5531 | Jan 6 |
| Crucifixion | VII 27 5534 | Mar 23 |
| Resurrection | VII 29 | Mar 25 |
| Ascension | IX 8 | May 3 |

**11:** *BM 815* 17b I, 20—II, 17
Duplicate: **13** 3; partial parallels: **14, 45; 22.**
l. 5: computing back from W 5574 (l. 12) gives W 5542 for Claudius 1.

[1] Graf, CAL II p. 349.
[2] Cf. e. g., Hammerschmidt, Kalendertafeln, or Gelzer, Afric. I, p. 50; also **44** and Neugebauer [1981] p. 377.

*Table 20*

| | **6, 39** | | | **12, 134** | | | **8** | |
|---|---|---|---|---|---|---|---|---|
| | years | W | D | years | J | total | years | Σ |
| Adam→Flood | 2256 | | | | | | | |
| Flood→Tower | 544 | 2800 | | | | | | |
| Tower→Lord appears to Abraham | 640 | 3440 | | | | | | |
| Lord appears to Abraham→Moses the prophet | 440 | 3880 | | | | | | |
| Moses→birth of Christ | 1620 | 5500 | | | | | | 5500 |
| birth of Christ→years of Martyrs (era D) | 276 | | | | | | | |
| birth of Christ→conversion of Ethiopia | | | | 245 | | | 245 | 5745 |
| conversion→Diocletian | | | | 31 | 276 | 5776 | 31 | 5776 |
| Diocletian→Council of Nicaea, Constantine year 12 | 59 | | | 59 | 335 | 5835 | 41 | 5817 |
| Nicaea→Council of Constantinople | 58 | | 117 | 58 | 393 | 5893 = 11 C + 41 | 56 | 5873 |
| Constantinople→Council of Ephesus | 55 | | 172 | 55 | [448] | [5948] | 50 | 5923 |
| Ephesus→Council of Chalcedon | 21 | | 193 | [21] | 469 | 596[9] | 21 | 5944 |
| Chalcedon→Islam | 140 | | 333 | 170 | 639 | 6139 | 170 | 6114 |
| Beginning of the World→Islam | | 6109 | | | | | | |
| additional years | 891 | | | | | | | |
| completion | | 7000 | | | | | | |

continued in Ethiopic king-lists; cf. **57** and **56**

*Table 21*

| | | *Main Version* | | | | | *Variants* **1, 34** | | | |
|---|---|---|---|---|---|---|---|---|---|---|
| | | years | total | **7, 15, 19 1, 40, 46** | | 532ʸ-Cycles | | years | Σ | |
| 1. | Adam→birth of Noah | 1656 | | | | 3C+60 | Adam→Flood | 2128 | | 1. |
| | birth of Noah→Flood | 600 | 2256 | | | 4C+128 | Flood→Tower | 540 | 2668 | |
| | Flood→Tower | 571 | 2827 | | | 5C+167 | Tower→Abraham 75 | 1071 | 3739 | |
| | Tower→birth of Abraham | 501 | 3328 | | | 6C+136 | Abraham 75→Exodus | 430 | 4169 | |
| 5. | Abraham→Moses | 425 | 3753 | **46** | | 7C+29 | Exodus→Temple | 440 | 4609 | 5. |
| | Moses→David | 694 | 4447 | 703 | 4456 | 8C+191 | Temple→Captivity | 321/4 | 4930/33 | |
| | David→Nabukadanaṣor | 469 | 4916 | 460 | 4916 | 9C+128 | Captivity→Ezra | 70 | 5000/3 | |
| | Nabukadanaṣor→Alexander | 265 | 5181 | | | +393 | Ezra→birth of Christ | 536/30 | 5536/33 | |
| | Alexander→birth of Christ | 319 | 5500 | | | 10C+180 | birth of Christ→Martyrs | 316 | 5852/49 | |
| 10. | birth of Christ→conversion of Ethiopia | 245 | 5745 | | | +425 | Martyrs→Christianity | 111/4 | 5963 | 10. |
| | conversion→Diocletian | 31 | 5776 | **15, 40, 46** | | | Christianity→Gabra | 421 | 6384 = | |
| | Diocletian→Nicaea | 59 | 5835 | 41 | 5817 | +515 | Masqal | | 12C | |
| | Nicaea→Constantinople | 58 | 5893 | 56 | [5873] | | no continuations into Ethiopic king-lists | | | |
| | Constantinople→Ephesus | | | 50 | [5923] | | | | | |
| 15. | Ephesus→Chalcedon | | | 21 | 5944 | | | | | |
| | Chalcedon→Islam | 170 | 6114 | | | | | | | |
| | Nicaea→Gabra Masqal | 94 | 5929 | | | 11C+77 | | | | |

continued in Ethiopic king-lists

**12:** *BM 815* $17^b$ II, 17—$18^b$ I, 1
Duplicate: **134**; partial parallel: **8**; cf. Table 20 (p. 133). For the time after Christ cf. also **6** and **39.** Many additional information (mainly theological) about the councils. Cf. Weld Blundell, p. 498. Nicaea: date in l. 3 Constantine 12; in **52** l. 6 Constantine 10; correct would be Constantine 13; cf. also p. 54.
l. 6: 596[9]. The text has only 5960.

**131:** *BM 827* $120^a$ I, 16—$120^b$ I, 9
Duplicate: **9.** For the colophon ($120^a$ I, 1—16) cf. **9.**
Parallels: **5, 30, 38; 31, 32; 48, 50.** Cf. Table 1 (p. 29) and Table 21 (p. 134).

**132:** *BM 827* $120^b$ I, 9—II, 3
Duplicate: **10,** q. v.

**133:** *BM 827* $120^b$ II, 4—$121^a$ I, 3
Duplicate: **11**; partial parallels: **14, 45; 22.**

**134:** *BM 827* $120^a$ I, 3—$121^b$ II, 7
Duplicate: **12**; partial parallel: **8**; cf. Table 20 (p. 133). For the councils, cf. above **12.**
l. 6: 596[9]; the text has only 5960.

**14:** *BMA 16217* $13^a$ I, 3—13
Almost duplicate: **45**; partial parallels: **11, 133; 22.**
l. 14: omitting Claudius $14^y$.

**15:** *BMA 16217* $19^a$ I, 2—$19^b$ I, 4
Duplicates: **40, 46**; near Duplicate: **2**; cf. Table 21 (p. 134).

**16:** *BMA 16217* $20^a$ I, 4—II, 11
Parallels: **27, 41.**
l. 8 and 9: inverse order; cf. Table 2 (p. 32).

**17:** *BMA 16217* $20^a$ II, 11—$21^a$ I, 12
Parallels: **24, 28, 42**; cf. Table 4 (p. 35).
l. 18: reading 7 certain.

**18**: *BMA 16217* 21[a] I, 12—II, 18
First half of the following parallels: **25, 29, 43**; cf. Table 6 (p. 38).
l. 5: ʽIyorbeʼam, a king of Israel; cf. **26** l. 1.
l. 7: 41 could also be read 44 or 49.

**19**1: *BMA 24995*, 30[b] II, 1—31[a] I, 4
Duplicates: **31** (with cycles), **32**. Cf. Table 1 (p. 29).
Parallels: **1, 34; 7, 15, 40, 46**. Cf. Table 21 (p. 134).

**19**2: *BMA 24995* 32[a] I, 12—25
The only date that occurs also in other tables is 5536 (in l. 10) for the Birth of Christ (cf. **1, 2, 33, 34, 35**). The date 2108 for the Flood (l. 3) is reminiscent of 2128 in **1** and **34**.
l. 8, 9: Cleopatra: kâleʽe badarâ.

**20**: *BN 160 A* 2[a], 1—11
The Persians, cf. p. 44.
l. 20: 276, error for 176.

**21**: *BN 160 C* 2[a], 20—28
The Ptolemies; cf. p. 48.
l. 3: Argob, or Akâb, or Arsâb.

**22**: *BN 160 D* 2[a], 28—40
The Romans; cf. p. 48.
Partial parallels: **11, 13**3; **14, 45.**
l. 4: Kanun 25, error for Kanun 29.
l. 5: 5569, error for 5669.
l. 6: 34 or 31.
l. 10: Ḥedar (III)as well as the date 25 make no sense.
l. 12: 5563: expected 5703 (cf. above p. 53).

**23**: *BN 160*, 16[a] I, 13—II, 8; 80[b] I, 17—81[a] I, 5
Duplicate versions, excepting the two last entries (which I do not understand; both totals are incorrect).

l. 5: Days and hours for the Capture of Jerusalem could suggest an astrological background; cf. also **25** and **29.**

l. 7: I do not know of "weeks" of 62 years excepting here and in **29** l. 26.

l. 9: The year 5496 as year of the Resurrection may have connections with the "Spanish era" (cf. Neugebauer [1981] p. 377).

**24**: *BN 160*, $76^a$ I, 1—$76^b$ II, 8
Introductory: $76^a$ I, 1—13
l. 1: 6257 error for 2257, i. e. Noah 201; cf. above p. 131 note to **5.** Cf. also Neugebauer [1981] p. 372, Hilarianus.
Main part: $76^a$ I, 14—$76^b$ II, 8: close parallels: **17, 28, 42;** cf. Table 4 (p. 35).
l. 18: [6] restored from **28** or **42.**

**25**: *BN 160*, $76^b$, II, 8—$77^a$ II, 5
Parallels: partial **18; 29, 43.** Cf. Table 6 (p. 38).
l. 22, 23: throne names.
l. 24: the actual total would be $516^y$ $7^m$ $4^d$ $3^h$.

**26**: *BN 160* $77^b$ II, 5—$77^b$ I, 19 (margin)
Kings of Judah, Israel, Edom; cf. Table 7 (p. 40).
l. 18: the actual total would be $235^y$ $6^m$ $30^d$ (i. e. $7^m$).

**27**: *BN 160* $77^b$ II, 1—$78^b$ II, 2
Parallels: **16, 41.**
l. 2: 430 instead of 435; l. 3: 620 instead of 625; l. 4 [100]: restored from the total but hardly correct; l. 8: 1384 instead of 1381 (total correct); l. 9: 128 instead of 188 (total correct); l. 15: 130 corrected to 134 (as in **41** l. 16) but not used in the total; l. 22: cf. **41** l. 27.

**28**: *BN 160* $78^b$ II, 2—$79^a$ II, 11
Parallels: **17, 24, 42;** cf. Table 4 (p. 35).
l. 0: 3790 from **27** l. 22.
l. 7: 3907 instead of 3997, upsetting all subsequent totals.

**29**: *BN 160* $79^a$ II, 12—$80^a$ I, 9
Parallels: partial **18; 25, 43.** Cf. Table 6 (p. 38).
l. 0: 4205 from **28** l. 24.
l. 10: 4409 instead of 4429 (isolated error); l. 21, l. 23: throne names;
l. 23: 4730, expected $4721 = 4205 + 516$; l. 26: 5500 instead of $5500 - 434 = 5066$; cf. **23** l. 7 for "weeks" of 62 years.

**30**: *E 215* $64^a$ II, 23—$64^b$ II, 28
Duplicates: **5, 38.**

Parallels: **9, 13** 1; **31, 32; 48, 50.** Cf. Table 1 (p. 29) and Table 21 (p. 134).
l. 8: Amram written ʿâmabarm or . . . r(hâ)m, the hâ above the line.
l. 11, 12: Ptolemy: ʾabṭel (or ʾabṭal-) mêwos.
l. 15: reading 6917 (=13C+1) certain; cf. **5** and **38.**

**31:** *E 215* 72[b] I,21—73[a] II,7
Duplicates: **32** (only 3C mentioned); **19** 1 (no cycles).
Parallels: **5, 30, 38; 9, 13** 1; **48, 50.** Cf. Table 1 (p. 29) and Table 21 (p. 134).
l. 14: 6300 instead of 6306; correct in **67** l. 3.

**32:** *E 215* 73[a] II,15—73[b] II,8
Duplicates: **31** (with all cycles), **19** 1 (no cycles).
Parallels: **5, 30, 38; 9, 13** 1, **48, 50.** Cf. Table 1 (p. 29) and Table 21 (p. 134).
Parallel to 37[b] I,15—II,8 (i. e. **32** l. 9—14 and **67**): *BN 64*, translated in Chaine, Chron. p. 112f.
l. 4: 4328 instead of 3328; l. 14: 5929 written 5000-90-120 wa 9.

**33:** *E 215* 73[b] II,27—74[a] I,23
Near duplicate: **47**; for col. [Σ] cf. **47.** Note that the majority of intervals ends in 0; cf. above p. 67 and **47** from 2000 to 5500.
l. 7: 235 instead of 245; l. 8: 182, in **64** 184; l. 10: 133 as in **65** l. 2.

**34:** *E 2063* 26[a] II,11—27[a] I,8
Duplicate: **1.**
Parallels: **7, 15, 40, 46; 19** 1; cf. Table 1 (p. 29) and Table 21 (p. 134).
l. 6, 8, 10: 324, 530, 114: correct version 321, 536, 111 in **1.**
l. 9: Martyrs, actually era G.
l. 12: Gabra Masqal = W 6384 = 12C (not 13C).

**35:** *E 2063* 27[a] I,9—II,16
Duplicates: **2, 52,** but cf. the variants shown p. 43.

**36:** *E 2063* 27[b] I,1—II,15
Duplicate: **3.**

**37:** *E 2063* 27[b] II,36—28[a] II,12
Duplicate: **4.**
l. 11: reading 5776 certain, error for 5777 (as in **4**).

**38:** *E 2063* $44^{a}$ II,5—$44^{b}$ II,16
Duplicates: **5, 30** — but no 1 to 7 at the end.
Parallels: **9, 13**1; **31, 32; 48, 50.** Cf. Table 1 (p. 29) and Table 21 (p. 134).

**39:** *E 2063* $46^{a}$ II,7—$47^{a}$ II,2
Duplicate: **6.** Cf. Table 20 (p. 133); also **8** and **12, 13**4.
l. 11: Chalcedon→Islam $140^{y}$ should be $170^{y}$, with the total 6114 (as in **8, 15, 40, 46**).

**40:** *E 2063* $47^{a}$ II,2—$47^{b}$ II,8
Duplicates: **7, 15, 46.**
Parallels: **1, 34; 19**1. Cf. Table 21 (p. 134) and Table 1 (p. 29).

**41:** *E 2077* $154^{b}$ I,17—$155^{a}$ I,6
Parallels: **16, 27.** Cf. Table 2 (p. 32).
l. 2: 972 error for 912.
l. 10: 1656: garbled writing 10-100-6 wa 150 wa 6 (i. e. 1006 and 156).
l. 27: cf. Exodus 12, 40; also Gelzer, Afric. I,86.

**42:** *E 2077* $155^{a}$ I,7—23
Parallels: **17, 24, 28.** Cf. Table 4 (p. 35).
l. 18: $397^{y}$ is the correct total of the numbers listed, i. e. from Kuesa to Somson 20. Hence Kuesa 1 is treated here as the date of the Exodus instead of as the end of the 40 years in the Desert.

**43:** *E 2077* $155^{a}$ I,23—II,13
Parallels: **18** incomplete, **25, 29.** Cf. Table 6 (p. 38).

**44:** *M.* Guidi [1896] p. 379 I,23—380 I,18 (without translation or commentary). No parallel in my material; in part Amharic.

**45:** *P 5884* $7^{b}$ III,10—20
Almost duplicate: **14.** Partial Parallels: **11, 13**3; **22.**

**46:** *P 5884* $12^{b}$ II,5—$13^{a}$ I,14
Duplicates: **7, 15, 40.**

Parallels: **1, 34; 19** 1, **31, 32**; cf. Table 21 (p. 134).
l. 6: 703, 4456 errors for 694, 4447.
l. 7: 460 error for 469.

**47**: *U 3* $63^{\mathrm{a}}$, 2—$63^{\mathrm{b}}$, 3.
Near duplicate: **33**; cf. also **31, 32.**

**48**: *V 1* $205^{\mathrm{a}}$ II,41—$205^{\mathrm{b}}$ II,11
Close parallel: **50**; different versions: **5, 30, 38; 9, 13** 1; **31, 32.** Cf. Table 1 (p. 29) and Table 21 (p. 134).
l. 5: 143: text, incorrectly, 543.
l. 11: lada, perhaps correction of Faluda to Falada. Meaning Philadelphus certain; (the text adds: "the second Ptolemy").
l. 12—15: Diocletian, error for era G.
l. 15: 6916: final 6 certain; expected: 7.

**49**: *V 1* $205^{\mathrm{b}}$ II,28—$206^{\mathrm{a}}$ I,20.
l. 7: restored from **53** l. 1.

**50**: *V 1* $207^{\mathrm{a}}$ I,17—II,1.
Close parallel: **48**; different versions: **5, 30, 38; 9, 13** 1; **31, 32.** Cf. Table 1 (p. 29) and Table 21 (p. 134).
The text nowhere mentions the 532-year cycle; cf., however, **48** and **5.** Beginning with line 8 the association of the names with the years becomes very insecure. The terms weṭen ("beginning") and ṣalṭ ("ending") occur, as far as I know, in no other chronological list.
l. 11: the text says "Ptolemaius Philadelphus, the second of the Ptolemies".

**51**: *V 1* $207^{\mathrm{a}}$ II,25—39.
No other text gives 2068 for the Flood; also 5562 for the birth of Christ is only attested here and in **52.**
l. 6:the restoration [434] is required by the continuation in **72.**

**52**: *V 1* $207^{\mathrm{b}}$ I,16—36.
Duplicates: **2, 35.**
For 5562 as the year of the birth of Christ, cf. **51.**

**53**: *V 1* $204^{a}$ II,39—$205^{a}$ II,8.
Cf. also Table 5 (p. 37).

l. 1: the birth of Lewi is also mentioned in **49** but without a date. From the present text it would follow: from the birth of Jacob to the birth of Lewi 3556—3473 = $83^{y}$.

l. 7: Gâdeyon is given $40^{y}$ in **24** and in **28**.

l. 8: Yeftâḥêl 6 (reading 6 is certain): 6 is also possible in **42** but 7 is certain in **17**.

l. 9: Somson $20^{y}$: same in **24, 28, 42.**

l. 10: Samuel $20^{y}$: in **17** "Sâmu'êl, the prophet, 22"; 85: error for 80.

l. 12: Olympiad 1, 1 = W 4745. Cf. above p. 43.

l. 18: D 76 = 11C = G 0.

**54**: *Vi 6* $2^{b}$, 17—26.
No parallels in the present material; but cf. **4** and **44.**

| | | | |
|---|---|---|---|
| Possible emendations: | l. 6 | 216 | 5181 |
| | 7 | 279 | 5460 |
| | 8 | | 5500 Christ |
| | 9 | | 6114 Islam |
| | 10 | | 6705 Takla Hâymânot. |

Addition of $456^{y}$ to 6705 gives a more plausible date, 7161, for Takla Hâymânot; cf. 7171 in **58.**

*Ethiopic Kings (***55** *to* **72***)*

**55**: *B 84* $22^{b}$ II,12—$23^{a}$ I,26; continued from **7.**
The date 6306 (l. 3) for Yekuno Amlâk is $456^{y}$ earlier than the historically plausible date W 6762. All subsequent dates show the same difference. Cf. for this "reduced" chronology above p. 56.

**56**: *BM 754* $4^{a}$ I,9—II,11; continued from **8.**
Duplicate: **68**; parallels **55** and **58.**

l. 17 to 20: the text gives here the throne-names instead of the civil names in **68.**

l. 21: $10^{y}$, but $9^{y}$ in **68**; cf. also 9 + 1 in **60** l. 15 and 16.

**57** 1: *BM 815* $18^{b}$ I,11—$19^{a}$ I,14; continued from **12.**
Duplicate: **57** 2.

l. 1: the continuation from **12** l. 6 and 7 would require:

| | | |
|---|---|---|
| Chalcedon | | 5969 |
| Chalc.→Islam | 170 | 6139 |
| Islam→Yek. Am. | 623 | 6762 |

l. 6: Weld Blundell: Yom instead of Wadem.
l. 13: Weld Blundell incorrectly 17 instead of 16.
l. 15: from Yekuno Amlâk (l. 2: 6762) to the completion of 13C=6916 elapsed 154ʸ; the text has incorrectly 151.
l. 33: 24ʸ added by later hand.

**57**2: *BM 827* 121ᵇ II,1—122 II,11; continued from **13**4.
Duplicate: **57**1.
l. 15, 16: cf. note to **57**1 l. 15, 16.
l. 32: throne name A'elâf Segad; cf. **57**1 l. 32.
l. 34: not in **57**1.

**58**: *BMA 16217* 19ᵇ I,2—20ᵃ I,4; continued from **15**.
Partial parallel: **70**.
l. 1: Islam, error for Adam; 6662 error for 6762.
l. 5: the total 112 includes the 29ʸ which were omitted between Yekuno Amlâk and Amdâ Ṣeyon (correct in **57**1 l. 4 to 7). The same correction is applied to col. W.
l. 10: 49 error for 34 (as in **57**1 l. 17); cf. also note to l. 13.
l. 12: between the lines: "7000 years, i. e. Lebra Dengel".
l. 13: 227 error for 127. This total is based on the correct value 34 in l. 10.
l. 23: 7171; cf. **54** l. 10.
l. 29: W 7257 for the last year of Io'as corresponds to A. D. 1764/5 (instead of 1769); 40ʸ written by later hand above the line.

**59**: *BMA 16217* 21ᵇ I,1—17
Written, according to the colophon (I,17—II,4), in A. D. 1796.
Because of water damage in part illegible.
No parallel known to me.
l. 2: **61** *A* l. 52 gives 1809ʸ for the kings of Aksum from Arwê to Bâzên 8.
l. 3: Arbeḥa, probably misspelling of Abreha; **62** *B* l. 13 counts 13 kings, **63** *D* l. 5 and *E* l. 19 count 19 kings.
l. 4: **64** *F* counts 9 kings and 184ʸ for A. and A. to Gabra Masqal.

l. 5: Anbas Wedem, perhaps Wedem Ârâd; cf. also **64** *F* l. 22.
l. 8: in Dillmann [1853] p. 351 Takla Hâymânot is given the title Abuna. In **58** his date is given as 7171, in **57** 2 as 7201.

**60**: *BMA 24995* 31ª I, 1—II, 17; continued from **19** 1.
l. 2 to 6: cf. **71** l. 4 to 7.
l. 7: 34 written like 20 wa 7 (?).
l. 9: added: "peace upon him; he changed his allegiance (to Catholicism) and expelled the Franks (i. e., the Portuguese)".

**61**: *BN 160 A* 7ᵇ, 1—20; *A´* 90ª II, 6—14 (followed by a list of Aksumite patriarchs).
Parallels: Dillmann [1853] p. 341—344; Conti Rossini [1909] p. 286 (No. 1)—291 (No. 48).
Kings of Aksum; cf. above p. 58 and **59** l. 2: Arwê to Bâzên 25 kings. Dillmann p. 341, 2 A: 21 kings. Conti Rossini p. 289: 27 kings.
l. 5: Sad, sic for Sâdo (cf. Conti Rossini p. 286) where Sabâṣ (or Sabaḍa, not Sêbado) is said to have resided for 50 years.
l. 6, l. 11: mikidâ (or similar): Queen of Sheba; cf. Kebra Nagast, passim.
l. 12: Leḥkam, i. e. Menelik; Kebra Nagast ch. 32.
l. 17: Dillmann, and following him, Conti Rossini, take 'eska qatr literally, meaning "until noon", and assign consequently Qâsyo a reign of half a day. This is not only absurd (and without parallel in our material) but ignores what follows: zamâwaṭ 60 wa 4. Instead Dillmann (p. 341 A 7) gives to "Mawaṭ" $8\frac{1}{3}$ years (which is his way of rendering $8^{y}$ $4^{m}$) while Conti Rossini has (p. 287 No. 13) "mawaṭ (var. awṭeṭ) 8 ans et 1 mois" where the 1 can be taken as a misreading of 4. Our text has clearly 60 wa 4. Now 60 can be easily misread as 8; hence the most plausible interpretation of this entry is "(from) Qâsyo to Qatr of Mawaṭ 64 (years)".
l. 48: corresponds to Conti Rossini p. 291 No. 48.

**62**: *BN 160 B* 7ᵇ, 20—25
Parallels: Dillmann [1853] p. 344, 21—31; Conti Rossini [1909] p. 291 No. 49—p. 292 No. 59—60.
The total of the reigns listed in the first 11 lines is 193 (+x) years.

**62**: *BN 160 C* $7^b$, 25—30
Parallels: Dillmann [1853] p. 346/7, 3A; Conti Rossini [1909] p. 292 No. 59—p. 295 No. 91; partial parallel: **72**, ending with Ḍeḥam 28 and a total of 6177 years. Cf. Table 22 (p. 145) and p. 65.
l. 1: Abreha $10^y$, probably an error for $12^y$ (as in Dillmann and in **72**).

**63**: *BN 160 D* and *E* $7^b$, 30—$8^a$, 21
Parallels, to **63** *E* only: Dillmann [1853] 2B, 2C, p. 345/6; Conti Rossini [1909] p. 269 No. 1—p. 271 No. 15—16.
*D* l. 6: 2104, error for 1104.
l. 7: the 426th year of the 11th cycle is the year 5320 + 426 = W 5746 = J 246; i. e. the first year after the conversion of Ethiopia. This date has nothing to do with the construction of the Temple.
*E* l. 1: from here (until *G* l. 10) closely parallel to Kebra Nagast (Bezold p. 173, trsl. p. 138).
l. 8: K. Nag.: Sarguây.
l. 14: K. Nag.: Mâ'albâgâd or Mâ'albâr.

**64**: *BN 160 F* $8^a$, 21—28
Parallels: Dillmann [1853] 3B p. 347—349; Conti Rossini [1909] p. 271 No. 17—p. 274 No. 47.
l. 10: K. Nag.: Bazagâr.
l. 20: K. Nag.: only Mâ'edây.
l. 21—23: K. Nag.: only Kalâwdewos.
l. 26: K. Nag.: Zamaz.
l. 35: Gabra Masqal→Delna'od $244^y$; cf. **67** l. 2 and **71** l. 3: Gabra Masqal→Zague $244^y$.

**65**: *BN 160 G* $8^a$, 29—36
The date 6400 for Yesḥaq (l. 17) is $456^y$ earlier than the correct date 6856 which happens not to be attested in the present material.
l. 2: Zague→Yekuno Amlâk: $133^y$; the same in **67** l. 3.
l. 10: here ends Kebra Nagast.

**66 1**: *BN 160* $16^a$ II, 19—$16^b$ II, 9
Duplicate: **66 2**; cf. also **71**.
l. 2: 428 on the margin, 8 not certain.

*Table 22*

| | 62 *C* | | Dillmann [1853] p. 346/7 No. 1—29 | | | Conti Rossini [1909] p. 292/3 No. 59—91 | | | 72 | |
|---|---|---|---|---|---|---|---|---|---|---|
| | | | | | Variants | | | Vari-ants | | |
| 1. | Abreha | 10 | Aṣbeḥa 'ela Abreḥa | 12 | | Abrehā-Aṣbeḥa | | $27^{y}$ $6^{m}$/ $12^{y}$ | Abrehā | 12 |
| | Asfeḥa | 3 | Asfeḥa | 7 | | Asfeḥa | 7 | 3 | Af'seḥa | 7 |
| | Śāhl | 14 | Śahl | 14 | | Sāhel | 14 | 17 | Śāhl | 14 |
| | | | Adẖanā | 14 | | Adẖana | 14 | | Adẖanor | 14 |
| 5. | Rete'e | 1 | Rete'e | 1 | | Rete'e | 1 | 4 | Reta'e | 1 |
| | Ēsfeḥ | 5 | Aṣbeḥa | 5 | 16 or 17 | Asfeḥ | 1 | 5 | Asfeḥa | 5 |
| | Aṣbeḥa | 16 | | | | Aṣbeḥa | 5 | | Aṣboḥ | 10? |
| | Amināda | 7 | Amidā | 16 or 17 | 6 or 7 | Amēda | 16 | 7 | Amidā | 7 |
| | Abrăẖ | $2^{m}$ | Abrehā | $6^{m}$ | $2^{m}$ | Abrehā | $6^{m}$ | | | |
| 10. | Śahl | $2^{m}$ | Śāhl | $2^{m}$ | | Sāhel | $2^{m}$ | 2 | | |
| | Gabaz | 14 | Gabaz | 2 | 14 | Gabaz | 2 | 3 | Gabaz | 14 |
| | Sehul | 4 | Sehul | 1 | | Seḥul | 1 | 4 | Śehu'ul | 10 |
| | Aṣbeḥa | 3 | Aṣbaḥ | 3 | 2 | Aṣbaḥ | 3 | | Aṣbuha | 3 |
| | Abreẖa-Eder | 17 | Abreh-Adẖanā | 16 | | Abreh-Adẖānā | 16 | | Abrehā | 17 |
| 15. | Ṣaḥam | 18 | Ṣaḥam | 28 | | Ṣaḥam | 28 | 18 | Ḍeḥam | 28 = 6177 |
| | Amidāb | 12 | Amidā | 12 | — | Amidā | 12 | 17 | | |
| | continued to Constantinus | | continued to Constantinus and Gabra Masqal | | | | | | | |

l. 3: 663: 3 not certain, corrected from a 5.
l. 4: referring to the Zague Dynasty.
l. 6: 984: sic for 987 (as in **66** 2 l. 8).
l. 13: on the margin 70 or 73; referring to 33 or to 1298.

**66** 2: *BN 160* $80^{a}$ I, 9—$80^{b}$ I, 16
Duplicate: **66** 1; cf. also **71**.
l. 4: Gebra written 'agabra.
l. 15/17: G 206 = W 6590 = 12 · 532 + 206.

**67**: *E 215* $73^{b}$ II, 3—37; continued from **32**.
Parallel: *BN 64* $58^{b}$—$60^{b}$, translated in Chaine, Chron. p. 112f.; misprint in l. 3: 138 for 133.
l. 1: Kâhen for Kâleb; garbled writing for 5929: 500 - 90 - 120 wa 9.
l. 4: the text has 6080 instead of 6380.
l. 6: text: 6421 for 6471.

**68**: *E 2063* $47^{b}$ I, 7—II, 7; continued from **40**.
Duplicate: **56**.

**69**: *E 2077* $155^{a}$ II, 21—III, 31; continued from **43**.
l. 1: Zague→Yekuno Amlâk written on the upper margin of col. III; $336^{y}$: cf. **60** l. 3 and **65** l. 2 which give $133^{y}$.
l. 41: [ left blank or written in red (illegible).
The total of the data given in our text would be $509^{y}$ $9^{m}$ $2^{d}$; cf. Chaine, Chron. p. 246/7: 7262—6762 = $500^{y}$.

**70**: *P 5884* $13^{a}$ I, 10—III, 20; continued from **46**. Beginning with l. 24: later hand.
Partial parallel of **58**.
l. 14: Chaine, Chron. p. 247/8 names some 20 kings in this interval of $34^{y}$.

**71**: *U 3* $63^{b}$, 3—$64^{a}$, 16; continued from **47**.
Parallels: **66** 1, **66** 2.

**72**: *V 1* $207^{a}$ II, 39—$207^{b}$ I, 9; continued from **51**. For the period from Christ to Conversion, cf. **62** *B*.
Written in a very crude hand.
Parallel: Conti Rossini [1909] p. 292 No. 59, 60—p. 294 No. 75. Cf. also Table 22 (p. 145).

l. 1: the commonly accepted date of the Conversion is 5745 (=5500+245), not 6035.

l. 8 and 10: meaning of ga (or bâ$^?$) unknown.

The 14 reigns listed would total $142^y$ if one accepts the readings suggested here. This would give for Ḍeḥam 28 the proper date 6177; cf. above p. 65 and Table 22 (p. 145).

For 'ela cf. p. 65/66.

## SUBJECT INDEX

Abraham 18, 31, 36, 41, 59 n. 58
Abreha and Aṣbeḥa ("A. and A.") 59, 61, 62, 66
Abu Shaker 28, 33, 52 n. 36, 116
Abuna 143
A. D. 8, 30, 52
Aksum *see* kings
Alexander 19, 46, 105
Amharic 127, 139
Amidâ 62, 66, 71, 98, 109, 145
Annianus 30
Apocalypse, Daniel 46
Arabic calendar *see* Islam
Arwê (King Snake) 58, 59, 60, 61, 142, 143
Aṣbeḥa *see* Abreha
astrology 39, 136
Augustus 20, 50, 53, 89, 105

Babylon *see* Captivity, *see* Tower
Babylonian kings 44
Baltasar 44 n. 26
baryodes (cycle) 27, 96, 100
Bâzên, king 59, 60, 61, 142
Bible 23
Bicornute 46
Bizan 56
BN 160 44, 48, 53, 58, 59
Bodleian MSS 53
Byzantine chronography 27, 30, 33, 46

Captivity in Babylon 19, 44, 108, 110
in Persia 44, 45, 46
Chalcedon, council 21, 54, 133, 134, 139, 142
Christ 20, 23, 29, 41, 42, 53, 55, 57, 105, 133, 134
chronology, reduction by 456[y] 58, 111
Claudius 49, 50, 76, 79, 132, 135
Cleopatra 19, 48, 50
Computus *see* Coptic
Constantine 51, 53, 61, 135
Constantinople, council 21, 28 n. 6, 54, 57, 111
conversion of Ethiopia 20, 58, 61, 96, 97, 134, 144, 146, 147
Coptic computus 27, 72, 99, 131
councils 21, 54, 57, 80, 82, 100, 134, 135
Cycles, 532 years 8, 22, 27, 30, 73, 75, 77, 86, 96, 98, 100, 106, 107, 109, 113, 129, 130

Daniel, Apocalypse 46
David 38, 41, 59, 134
Debra Bizan *see* Bizan
Ḍeḥam *see* Ṣaḥam
Deluge *see* Flood
desert, wandering after Exodus 33, 109, 139
Dillmann 58
Diocletian 6, 23, 29, 41, 109
Diocletian era (*see also* G) 28, 134

Earthquake 124
Easter-cycle 27
Edom *see* kings
'ela, za'ela 65, 66, 119, 130
Embarm 29, 30 n. 8, 93, 131
Ephesus, council 21, 54
eras (*see also* Diocletian, G, World) 8, 28, 30
Ethiopic kings 55, 111, 141
Eustatios 127